Consilience, Truth and the Mind of God

Richard J. Di Rocco

Consilience, Truth and the Mind of God

Science, Philosophy and Theology in the Search for Ultimate Meaning

With Contributions by Arthur J. Kyriazis and Edgar E. Coons

 Springer

Richard J. Di Rocco
Psychology Department
University of Pennsylvania
Philadelphia, PA, USA

Psychology Department
St. Joseph's University
Philadelphia, PA, USA

ISBN 978-3-030-01868-9   ISBN 978-3-030-01869-6   (eBook)
https://doi.org/10.1007/978-3-030-01869-6

Library of Congress Control Number: 2018959105

This Springer imprint is published by the registered company Springer Nature Switzerland AG
The registered company address is: Gewerbestrasse 11, 6330 Cham, Switzerland

If natural Philosophy in all its Parts, by pursuing this Method, shall at length be perfected, the Bounds of Moral Philosophy will be also enlarged. For so far as we can know by natural Philosophy what is the first Cause, what Power he has over us, and what Benefits we receive from him, so far our Duty towards him, as well as that towards one another, will appear to us by the Light of Nature.

(Sir Isaac Newton, Opticks)

The most beautiful and most profound experience is the sensation of the mystical. It is the sower of all true science. He to whom this emotion is a stranger, who can no longer wonder and stand rapt in awe, is as good as dead. To know that what is impenetrable to us really exists, manifesting itself as the highest wisdom and the most radiant beauty which our dull faculties can comprehend only in their primitive forms – this knowledge, this feeling is at the center of true religiousness.

(Albert Einstein, The Merging of Spirit and Science)

*This book is dedicated to my father,
Mario Di Rocco, who set an example
of passion in the pursuit of wisdom.
His love of nature and learning is passed
on to others in honor of his memory.*

Preface

Since the dawn of human self-awareness, people have lived with a sense of existential dread that arises from the fear of death and more generally from fear of the unknown. Questions related to the origin of the universe and the purpose and meaning of human life, mind, and death are still a cause of curiosity, wonder, and fear. Is there an essential truth that is the essence and the source of all reality and being? Our human ancestors first attempted to explain reality by the invention of myths that attributed the control of nature's forces to gods or spirits with supernatural powers. Primitive religious beliefs were generated on this basis, and people felt a deep connection to the postulated but unseen realm of spirits. This same type of magical thinking is observed in the explanations that children offer for things they do not understand, including their own behavior at times. Is belief in God no more than this?

Many scientists would answer *yes* that belief in God is a mere superstition. They believe that the power of science to explain the wonders of the universe obviates the necessity of a supreme intelligence as its creative source. The idea expressed by these scientists is that humanity has had recourse to the concept of a Divine Creator as an explanation for the natural world because of large gaps in knowledge before the era of modern science. As science made progress in explaining nature, so the argument goes, the need for the so-called *God of the gaps* to rationalize the unknown became an obsolete crutch. This book explains why this specious argument is false by demonstrating that it is not by the narrowing of the gaps in knowledge that God is disproved by science but rather that it is within the edifice of the scientific understanding of physics, cosmology, biology, psychology, and philosophy[1] that God can be found. For it is in the natural history of the universe that we can find the chain of causative events that lead from the Big Bang to the origin of life and ultimately to

---

[1]The scientific understanding of philosophy is available to us in the understanding of brain mechanisms and behaviors responsible for deductive and inductive reasoning. The foundation of this understanding will be laid in Chap. 6 (Paleopsychology) and will be developed further in Chap. 7 (Mind Knowing Truth).

intelligent human mind that is competent to ponder this question and to demonstrate the necessity of God's existence, if indeed God is real.

The imperative to explain reality is a pervasive human trait that has led to the creation of the modern world religions, philosophy, and science based on *rational empiricism*.[2] Unfortunately, while they both seek explanations of reality, the dialogue between science and religion has polarized dramatically, especially with regard to the question of God's existence. In the context of this polarization, a number of irrelevant issues have been raised concerning the existence of God. These proxy issues provide "straw men" that are easily dismissed. One of these straw men, which concerns the literal truth of the Bible, has been accepted by both atheists and theists as a proxy issue for the existence of God. Atheists point to the success of science in generating knowledge that contradicts the literal interpretation of the biblical narrative, including its many allegorical passages. The debate in the seventeenth century between the Church and Galileo and the nineteenth century debate concerning the validity of Darwinian evolution provide two well-known examples. Many believers have taken up the diversionary challenge and have proposed various pseudoscientific arguments that are advanced under the banner of *creationism* or *intelligent design* to counter refutations of the Bible as literal truth. It is clear on the basis of prima facie evidence, however, that the existence of God does not depend on the literal truth of the Bible. It is likewise clear that the existence of God cannot be refuted by the efficacy of science as a method to discover truth about nature. The idea that the reality of God's existence is somehow "squeezed out" of scientifically narrowing gaps in human knowledge is false.

Consilience, Truth and the Mind of God is concerned with scientific, philosophical, and theological subject matter that I hope will contribute to an inquiry into the big questions that have puzzled humanity from the dawn of *sapience*.[3] While such a broad purview of subject matter may seem ambitious, the multidisciplinary approach is necessitated by the magnitude of the task at hand. In part, this book will consider whether a convincing argument for the existence of a unique, singular, and eternal supreme intelligence, commonly referred to as God, can be attained. Owing to the importance of the question, and the inherent difficulty to be expected in the search for the answer, it seems reasonable to expect that success will be facilitated by a broad approach rather than a more limited tack.

Moreover, from the perspective of consilience,[4] which posits the unitary nature and coherence of all knowledge, the perception of boundaries between disciplines seems more artificial than real and is likely to result primarily from the imperfection of understanding. This is to say that the distinctions that we perceive among the truths revealed in different areas of inquiry are essentially superficial. A corollary of this supposition is that deeper understanding of what at first may appear to be

---

[2]*Rational empiricism* is the modern scientific method established by Sir Francis Bacon and others.

[3]*Sapience* is the faculty of intelligence that is possessed by the modern human mind.

[4]*Consilience* refers to the inherent unity of all the different areas of knowledge. This concept will be explained in the chapters that follow.

unrelated truths should increasingly reveal elements of correspondence. From this perspective, it is fair to say that all truth has the potential to attain to a demonstration of God's existence, if God exists; but there's the problem, for that is what is to be proved. Therein also may be found a potential criticism of the book. There is the danger that the synthesis required is too encompassing and that it could be perceived as superficial as a result. Yet, there is a grand coherence among all that is or can ever be known that is suggested by the concept of *consilience.* So, while it may be necessary to consider information from diverse disciplines that may seem unrelated or perhaps irrelevant at first, consilience assures us that a synthesis is in fact possible. I hope that readers will find that this book provides at least the beginning of such a synthesis, whatever its flaws. To the degree that this is accomplished, it will become increasingly clear that a comprehensive approach is warranted in the search for ultimate meaning. I can say this much now; if you are seeking, keep reading. You will then judge for yourself whether what you find justifies your effort.

Many arguments have been offered for the existence of God over the millennia. Each has been criticized on various grounds. These will be reviewed, but the main philosophical argument that will be offered in this book will build upon a proof referred to as the Argument from Truth. According to the *Handbook of Christian Apologetics* by Peter Kreeft and Ronald Tacelli, this argument originates from St. Augustine and proceeds along the following lines. I have placed questions or areas that need further clarification in parentheses. The parenthetical questions indicate that the premises of Augustine's argument require further justification, which can only be found in epistemology, the theory of knowledge, as authors Kreeft and Tacelli emphasize:

Our limited minds can discover eternal truths about being (Why are these truths eternal?)

Truth properly resides in a mind (Why?)
    But the human mind is not eternal,
    Therefore there must exist an eternal mind in which these truths reside.

The authors continue*:*

And there is a good deal to be said for this. But that is just the problem. There is too much about the theory of knowledge that needs to be said before this could work as a persuasive demonstration.

A main premise of this book, upon which the argument presented in the final chapter relies, is that the theory of knowledge *is* in fact sufficient to support a more definitive and conclusive modified Argument from Truth for the existence of an eternal mind that is possessed by a transcendent, self-sufficient intelligence or mind that is the source of all existence. Along with the development of science, a sophisticated body of philosophical method has been achieved, both during the Classical period of the Greek philosophers and since the time of the Enlightenment. An argument is advanced herein that relies upon biologist E. O. Wilson's popular-ization of the idea of consilience, which has its origins in the classical philosophy of the Greeks and refers to the unitary nature and coherence of all knowledge, as well as upon the epistemology of John Dewy and Arthur Bentley, which they set forth in their landmark work, *Knowing and the Known*. Dewey and Bentley convincingly argue that knowledge, per se, has no independent existence apart from mind. Since truth is not created by humans, but can be discovered by them at any time or place, truth must exist before its discovery. This leads immediately to the fact that truth exists eternally. Since no finite human mind can apprehend or know the eternal consilient truth in its full extent, and truth has its only existence in mind, consilient truth must be apprehended by a *transcendent*[5] mind that has its eternal existence outside of space and time. That mind is the mind of God. If the premises of this *syllogistic*[6] argument are true, the conclusion necessarily follows. The alternative is that eternal truth exists in some ill-defined, mindless, ethereal Platonic *realm of ideals, ideas,* or *perfect forms*. The central argument in the final chapter of *Consilience, Truth and the Mind of God* is thus a modification of Augustine's Argument from Truth in which the concepts of consilience and the transactional aspect of knowing, as an act of mind, combine to support the existence of a potentially infinite and eternal corpus of truth that is defined as everything known about the infinite and eternal multiverse, its laws, and mathematics by the mind of the necessary eternal being, God, who is the supreme intelligence and source of existence.

Some practical guidance for the reader is in order. I tried to provide adequate explanation, and graphic illustration, for some of the more difficult concepts covered in this book. Chap. 3 covers some key concepts in physics that involve mathematical content at the level of high school algebra. Logarithms, probability, sigma notation, and the algebraic manipulation of equations are used to explain *entropy, information*,[7] and the relationship between these phenomena. If you are uncomfortable with the mathematical content, the examples and other information in the text should provide a reasonably sufficient basis for you to grasp the important ideas presented in Chap. 3. Footnotes provide important supplementary information, especially definitions of key terms, throughout the book, and the reader is urged to take advantage of the additional perspective that they offer.

Philadelphia, PA, USA   Richard J. Di Rocco

---

[5]In this context, *transcendence* refers to a state of being outside the bounds of space and time. Such a state of being must therefore be viewed as eternal, since it has no beginning or end in the sense that we understand such terms from our perspective within space-time. A "transcendent mind" is one that exists beyond the bounds of space and time. It is eternal.

[6]A *syllogism* is an argument that proceeds according to the rules of deductive reasoning, also referred to as Aristotelian logic.

[7]*Entropy* and *information* are terms from physics and mathematics that will be defined and examined in detail in Chap. 3 on Physics.

Acknowledgements

With the passing of years, I have grown increasingly mindful that if I have something to say, I had better do so. This book was written over the last few years, but a lifetime of curiosity and pondering has gone into its making. I especially would like to thank my wife, Rebecca, for her encouragement and support. I also want to thank my collaborator, Arthur Kyriazis, who not only made vital contributions to the essential Chap. 8 on arguments for the existence of God but also provided helpful commentary throughout the writing of the book. Likewise, I thank my thesis advisor and friend, Prof. Edgar (Ted) Coons of New York University. He not only helped to polish a student with some rough edges, and kept me moving forward despite my shortcomings, but also has been a lifelong intellectual collaborator. Much of the subject matter of this collaboration appears in our joint effort on Chaps. 5 and 6. I am also grateful to those who inspired my curiosity and commitment to lifelong learning. My father, Mario Di Rocco, introduced me to my first telescopic view of the heavens. He exemplified a sense of wonder in the contemplation of the universe and instilled in me the love of scholarship. I also want to thank Prof. Harvey Grill of the University of Pennsylvania, who has been a steadfast friend and counselor since our days together in graduate school. I owe thanks as well to many students who attended my courses over the years and to my children, from whom I have learned much. They serve as a constant reminder that hope for the enlightenment of humanity is justified by what I have observed in the minds and hearts of the young. In particular, I would like to thank one of my students, Abdullah Noaman, who read the final draft of the manuscript and provided helpful comments. Finally, I want to thank my mother who did the most for me by bringing me into this amazing reality in which we find ourselves. She is the one who taught me to expect more from myself and to discover and strive constantly toward the fulfillment of life's purpose. I thank you all.

Contents

Chapter 1
Introduction: The Search for Ultimate Meaning

Richard J. Di Rocco

There are more things in heaven and earth, Horatio, than are dreamt of in your philosophy.

William Shakespeare. Hamlet Scene I, Act V

*Who is this that darkens counsel by words without wisdom?
...Where were you when I laid the foundation of the world?*

Job 38: 2–4.

Je cherche a comprendre (The last intelligible words of the great French molecular biologist, Jacques Monod, which may be translated literally as, "I am searching for understanding" but which has the sense of, "I am trying to understand".)

Jacques Monod, in Judson H F (1996)

Abstract The modern human mind confronts many existential questions as it contemplates the vast external reality of the universe in which it exists, and the internal reality of its own perception, thought and self-awareness. Humanity is faced with many unanswered questions about the origin of the universe, life and mind. Most significantly, owing to our faculty of advanced intelligence, humans are confronted with the daunting implications of mortality, and this is the stuff of which existential crises are made. Defense mechanisms protect against the fear of the ego's annihilation in death, but fear of the ultimate unknown also provides motivation in the quest for understanding. This is inherently disquieting because it requires an admission of profound ignorance. Humility in the face of the unknown is essential, however, because without it the questions that lead to learning go unasked and unanswered. This chapter provides a broad overview of the scope of the

R. J. Di Rocco (✉)
Psychology Department, University of Pennsylvania, Philadelphia, PA, USA

Psychology Department, St. Joseph's University, Philadelphia, PA, USA
e-mail: richdi@upenn.edu

© Springer Nature Switzerland AG 2018
R. J. Di Rocco, *Consilience, Truth and the Mind of God*,
https://doi.org/10.1007/978-3-030-01869-6_1

1

epistemological, ontological and scientific questions that are addressed in succeeding chapters in the search for ultimate meaning.

Keywords Existential questions · Consciousness · Theory of Mind · Ontology · Ontology of truth

This book is intended to provoke the reader's consideration of fundamental existential questions that confront sapient beings in the universe: why does anything exist; does the existence of the universe necessarily imply the existence of a Creator; why am I here; where am I; indeed, what and who am I? Stand before a mirror to see your reflection. As a sapient human, you attribute mind to the being reflected in the image. You attribute mind to yourself. You are self-aware. You know that the mind behind the face reflected in the image, that perceives the image, is your "self". You are seeing yourself see yourself, much as one sees reverberating reflected images of images in the opposing mirrors of a hair-cutting salon. Fluidity of the attribution of body image and ownership has been demonstrated in recent virtual reality studies of embodiment. These fascinating experiments showed that it is possible to experience the localization of consciousness not in one's own head, but in a virtual child-like talking body (Tajadura-Jiménez A et al. 2017). The misattribution by an adult of body ownership, and hence the localization of consciousness, to the virtual body of a child was shown to have profound effects on perception of the size of objects, the pitch of the adult's speech and the adult's emotions. These experiments certainly give us reason to marvel at how the human brain provides the capacity for self-awareness and the localization of conscious experience within, and without, the body.

Or go to a high place far from city lights, and look up to see the Milky Way. Then reflect upon the fact that our galaxy of approximately 100–400 billion stars is only one of what has been estimated to be two trillion galaxies in the universe (Conselice CJ et al. 2016)! Even if we assume one trillion galaxies, and that on average these galaxies have only one billion stars each, this leads to a low estimate of one billion trillion stars in the known universe.[1] Contemplate the awesome reality of existence itself, and further that your mind evolved from inanimate matter as a result of a little less than 14 billion years of the natural history of the universe. You may then wonder, with some humility, *how* and *why*. Could one ever hope to have verifiable evidence of a phenomenon more radically amazing? Surely the creation of the universe, based on orderly laws of mathematics and physics from unknown antecedents, must be viewed as astounding by any standard! Similarly, the emergence of life and mind in the universe from non-living matter seems so improbable that it too is often viewed as miraculous, although there is no a priori reason or need to invoke supernatural mechanisms for these events that occurred after the universe and the laws of physics were established. The transition from inanimate matter to the first self-replicating molecules is not yet fully understood, but is justifiably presumed to

---

[1]In Chap. 4, "Cosmogenesis", we will consider the possibility that the known universe is merely one in an infinite number of universes that comprise a vast eternal network of existence that cosmologists call the eternal multiverse.

proceed on the basis of the laws of chemistry and physics upon which they rest. Questions also remain concerning the formation of the first cells, but the path from there to human life is better understood. Considering the 13.7 billion-year age of the universe, and the fact that the solar system is much younger at an age of approximately 5 billion years, the possibility that intelligent life has emerged on older stellar-planetary systems throughout the universe must be considered. The Milky Way alone contains at least 100 billion stars, so it is reasonable to entertain the hypothesis of extraterrestrial life somewhere in the universe despite how improbable life seems. The key event that remains unexplained by any means, however, is the origin of the universe.

Such experiences put consideration of existential questions concerning the nature of the universe and human consciousness in their proper context. These are truly profound and fundamental questions that are concerned with the means and meaning of existence, reality and our own consciousness. In the face of such questions one may ask why people not only take the universe or reality for granted most of the time, but also their own consciousness. Renee Descartes was a notable exception to the general tendency of humans to take consciousness for granted. He extolled his personal experience of consciousness as the foundation for any certainty he could have about truth and reality when he said *"Cogito ergo sum"*, "I think, therefore I am". This simple, yet profound, epistemological insight was important for Descartes, who was following a method of inquiry in which he doubted everything so that he could build an edifice of truth on the firm foundation of first principles only. Doubting everything he could know about reality soon led to a profound philosophical-existential crisis that was rooted in uncertainty about how he could trust his own thoughts, even the truth of his own existence. In this condition of self-doubt, however, he found his key insight that a thinking being must certainly *be*! Descartes thus assured himself, and the rest of us, that the self exists with epistemological certainty. This was the one thing he could know with certainty, the solid foundation or terra firma upon which he could build the rest of his philosophy, mathematics and science. According to *Theory of Mind*,[2] humans confer this certitude concerning *self* to each other. (Premack D G and Woodruff G 1978). *Cogito ergo cogitas* – "I think, therefore you think" is the essential premise of Theory of Mind.

What is the source of the universe? How can we explain the emergence of life and mind in the universe? Both science and religion grapple with these questions; and we do well to ponder them, because if nothing else they are humbling when properly understood. The humility thus obtained has great value, because it is a prerequisite

---

[2]David Premack has defined *Theory of Mind* as "the ability to attribute mental states—beliefs, intents, desires, pretending, knowledge etc.—to oneself and others and to understand that others have beliefs, desires and intentions that are different from one's own".

for honest inquiry. Every question is an implicit admission of ignorance,[3] which is inherently disconcerting. Yet without the awareness and acknowledgment that something is not understood, the questions that lead to learning remain unasked and unanswered. In the extreme, the pervasive willingness to take the universe, the existence of reality itself, as given begs the penultimate *ontological* question[4] that is the concern of scientific and religious inquiry alike. As with Mount Everest the universe is accepted because it is there, but is that where curiosity must end? Is there not a bit of arrogance in the acceptance of existence as an entitlement that does not require one to wonder, at least occasionally, *from where did it come*? Once the question is honestly confronted the absence of an answer is disquieting. To ask so great a question requires admission of profound, anxiety-provoking, ignorance. Anxiety of such magnitude is the likely cause of the pervasive and adaptive denial of the radically amazing nature of reality. Sapient beings could not function effectively for very long in a permanent state of awe or existential dread. Acceptance of reality as normal and given, therefore, may be viewed as an essential and profound example of the wisdom of the ego's defense mechanisms [See for example "The Wisdom of the Ego". (Vaillant G E 1993)]. Yet there are times when the ego's defenses break down, and this is the stuff of which existential crises are made. Even at the risk of evoking such angst, however, it becomes necessary to peer through the veil of the ego's defenses to see the matter for what it is and to confront the ultimate fear in the pursuit of wisdom. The ancient Greek philosopher, Parmenides was more than willing to take up the challenge (Burnet J 1920). Parmenides was persuaded that, because *nothing comes from nothing,*[5] *what is* exists eternally, is immutable, uniform and of one fundamental nature throughout. Parmenides thus anticipates quantum theory's ideas concerning the unitary nature of reality. Quantum entanglement, the notion that paired quantum entities such as electrons or photons and even large molecules are inextricably connected and instantaneously interactive over immense distances, has been shown to be an accurate description of reality. In addition, the notion of a multiverse, in which each universe such as our own gives rise to other universes in an infinite progression, raises the possibility of an eternal nature of reality and being that would validate the ontology propounded by Parmenides. We will consider these issues further in Chaps. 3 and 4 on physics and cosmology, and again in Chap. 8 in relation to arguments for the existence of God.

To the believer, faith in God is a gift. To the atheist, it is a manifestation of ignorance and superstition. Along the spectrum of opinion defined by these extremes, surely there is sufficient room for humility in the face of the unknown.

---

[3]*Ignorance* used in this context carries no pejorative significance. Rather, it designates the cognitive condition of a person in regard to something that is not understood. Ignorance does not define that person's intelligence. It only defines a lack of insight in relation to a question under consideration. Admission of ignorance, thus defined, is a manifestation of intelligence and intellectual honesty and is a necessary prerequisite for all learning.

[4]*Ontology* is the philosophy of being, existence and reality. The *penultimate question* concerning the origin of the universe leads naturally to the related *ultimate question* that concerns the existence of God as the source of created reality.

[5]*"ex nihilo, nihil fit"* – from nothing, nothing comes.

The universe poses the penultimate question. By its very existence it asks, and demands an answer to the question concerning itself, *why*. We must not be satisfied with the answer given by the mythical philosophy student of urban legend who, when faced with the one word question *why* on the final exam, earned a grade of A with the answer, *because*. We must go further to seek a more definitive and satisfying answer to the question. One must acknowledge in considering so monumental a question that a profound sense of humility is appropriate, and that the answer may require at least the possibility of a creative intelligence as the source of all being.[6] If not, then we must be resigned to living with ambiguity in regard to the cause of the universe, or multiverse as the case may be, and the question whether God is real.

The metaphysical-existential dilemma, and related fears that necessarily confront sapient beings, provide the motivation and the starting point in the quest to answer the ultimate ontological question concerning the existence of God. Quine calls the question *"What is there"* the Ontological Question. He states at the beginning of his essay titled, "On What There Is" (Quine WVO 1948):

> A curious thing about the ontological problem is its simplicity. It can be put in three Anglo-Saxon monosyllables: 'What is there'? It can be answered, moreover, in a word—'Everything'—and everyone will accept this answer as true. However, this is merely to say that there is what there is. There remains room for disagreement over cases; and so the issue has stayed alive down the centuries.

The most important of these cases concerns the existence of God. Perhaps a better ontological question than *what is there* would be, *why is there*. Why does anything exist? Jim Holt tackles this question in his book titled, "Why Does the World Exist?" (Holt J 2012). Holt takes on this question in his eminently readable, insightful book. He provides his own learned take, as well as impressions of his interviews with various luminaries in the fields of science, philosophy, metaphysics and theology as they react to the question, "why is there something rather than nothing". The related question *"how* can anything exist", has provoked extensive scientific inquiry into the story of the universe from the Big Bang to present time. Human understanding of the natural history of the universe has progressed to an advanced stage and may soon provide insight into the first moment of the Big Bang and its cause. This story is recounted in the chapters that follow. The crux of the matter may be stated as follows. Does the existence of reality, known and unknown, require a pre-existent or transcendent necessary being? If the answer to this most important of all questions is no, as atheists maintain, then we either must accept *nihilism*[7] as a way of being, or we require an alternative explanation not only for all of reality, but also for the existence of all of physical law, mathematical truth, indeed the existence of all truth.

---

[6]The case for such humility is made in the Book of Job, a biblical allegory in which the provocation of friends causes Job to seek a justification from God for the suffering he has been forced to endure. The answer that Job receives may serve as a reminder of the need for humility in the search for ultimate meaning: "Who is this that darkens counsel by words without wisdom? . . . Where were you when I laid the foundation of the world?" (Job 38: 2–4.)

[7]*Nihilism* is a philosophical argument that holds that nothing in the world has meaning or a real existence.

This in turn raises the question, *in what does truth exist, in what does it subsist and have its being*. We will see that the answer to this question provides a vital key to an argument for the existence of a self-sufficient supreme intelligence that is the cause of all reality.

This book attempts to provide an objective, rational basis for evaluating the proposition that God is real and that God is the cause of the universe in which we exist. Unfortunately, the dialogue that surrounds the question of God's existence is encumbered by confusion created by strident and polarizing arguments advanced by scientist-atheists on the one hand, and believers on the other. To progress beyond this impediment, it is necessary to examine the debate between these factions with a focus on highlighting factors that have led to unnecessary polarization, illogical arguments and confusion. Only having done this, will it be possible to find the common ground upon which religion and science rest and begin consideration of a more synthetic and productive line of inquiry that will lead to progress in understanding. An overview of the synthesis that is attempted in "Consilience, Truth and the Mind of God" is presented in the chapter outlines below. These provide a good sense of how the rest of the book not only covers the indicated chapter topics, but also how the information in each of them leads in a natural progression to the final synthesis in the search for ultimate meaning.

Chapter 2 examines the factors that have led to the historical polarization between proponents of science and religion. Theism and atheism are both based on faith because neither is competent to decide the question concerning the existence of God. In the absence of convincing arguments, proponents of each have polarized in the heat of debate. The essential question has been overshadowed by irrelevant proxy issues, such as whether the entire Bible is literally true or whether some of its passages are primarily allegorical. Issues such as this are irrelevant to the existence of God. I hope to show why the permanent polarization of science and religion is not inevitable, because it is reasonable to assume on the basis of consilience that as science and theology make progress in apprehending truth the statements they each make should increasingly correspond. Until that equivalence is revealed, agnosticism offers an honest starting point in the attempt to understand whether the universe, or multiverse, is the sufficient cause of itself or whether its cause derives from beyond its spatial and temporal bounds.

Chapter 3 provides a broad historical approach to relevant topics in physics that include the related concepts of entropy and information. The text provides adequate description of vital aspects of these areas of classical physics, so the reader can skip the equations as desired without much loss of understanding. The discussion of entropy and information lays a vital foundation for the understanding of life and mind that follows in later chapters. The main highlights of quantum theory, that point toward the underlying unitary and coherent nature of reality, follow the section on classical physics. The material on quantum physics provides an important foundation for the overview of cosmology in the next chapter, as well as a scientific context for arguments considered later in the book about the existence of a self-sufficient necessary being.

Chapter 4 provides an overview of what is known, or hypothesized, about the origin of the universe. The scientific approach to this problem is hampered by the fact that we can only examine the question retrospectively from our position within space-time. Physics takes us back to the Big Bang, but it does not allow us to see its transcendental cause directly. The achievements of physicists and cosmologists in regard to this daunting question are remarkable in light of this profound handicap. Science has not only succeeded in demonstrating extremely early events immediately after the Big Bang, but also has led to theoretical inferences about its origin by examining causal imprints that are accessible to science within space-time. The related concepts of inflationary cosmology and the eternal multiverse are of special theoretical significance because they seem to offer a path forward.

Chapter 5 explores the great mystery of abiogenesis, which concerns how life first arose on Earth from non-living matter; and how this happened in harmony with the laws of physics and chemistry. Of that much we can be certain, but of the processes of chemical evolution that led to the appearance of self-replicating organic molecules; and further, concerning how the mechanisms of metabolism and genetic transmission of information were encapsulated within the cell membrane, mystery remains. The question of abiogenesis defines the next great frontier in biology, and we must hope that its explorers will be as passionate and dedicated to the quest as were their scientific forebears who discovered the mechanisms of molecular biology when they attempted to answer the question, *what is life*. This question was posed by Erwin Schrödinger in his seminal book of the same title. While the mystery of abiogenesis remains, a great deal has been accomplished since then. This chapter will examine the basic mechanisms of molecular biology as it is understood today, as well as what we can infer about abiogenesis from first principles, and how all of this information informs the effort to discover the mechanisms of life's beginning.

Chapter 6 brings the concepts of entropy and information, that were developed in Chap. 3, to bear on the problem of the origin of life, and the evolution of living matter once it was established. The origin of human and animal memory mechanisms in the signal transduction mechanisms of pre-Cambrian single-celled organisms is described. This phenomenon offers one of the strongest examples of conserved cell and molecular mechanisms in biology. The subsequent rise of multicellular organisms during the Cambrian explosion approximately 500 million years ago is discussed in the context of predator-prey relationships that provided the selective pressure for the evolution of neural networks that were optimized for effective escape and predatory behaviors. The mechanism of learning in birds and mammals, is described as the prelude to understanding the emergence of the modern human mind with its amazing cognitive capabilities. The dawn of human meta-awareness in *Homo sapiens sapiens*, the human who knows he knows, is described and its implications for the emergence of existential fear is discussed. One of the consequences of increased intelligence, besides its ability to magnify fear of the unknown, is the corresponding compulsion to provide explanations for phenomena that are not understood. This characteristic of modern human thinking has been referred to as the *cognitive imperative*, and it is manifested in the magical thinking of early *Homo sapiens*, as well as contemporary children. Existential dread, the fear of annihilation of the ego at death, is discussed as the likely driving force for the emergence of mythology as proto-religious dogma that provided a sense of comfort and reduction of anxiety in our human ancestors, as well as in ourselves. A theory of the origin of evil in human behavior is offered, in which the impetus for genocide and global war can be found in the overvaluation by humans of the hypothetical constructs that we generate to explain the unknown. Humanity is the only species ready to kill to defend ideas, because we fear the loss of the presumed certainty and security that our constructs of reality provide.

Chapter 7 is concerned with the nature of truth, how it is discovered and proved. The examination of this issue begins by asking the question whether new mathematics is invented or discovered. It is somewhat surprising that mathematicians hold different opinions on this matter. Everyone will agree that *new mathematics* involves valid operations that lead to true statements or conclusions about the objects of those operations. So, the question concerning whether new mathematics is invented or discovered can be restated in an equivalent form; do humans invent or discover truth? When the question is posed this way, it is clear that truth is discovered but not invented. This follows from the fact that true statements of logic or mathematics can be proved using deductive reasoning that leads from a statement or theorem already known to be true to the statement which is to be proved. This is exactly what high school geometry students do when presented with a *given* truth, or axiom, and are asked to proceed from there to prove another statement, or theorem. The key point about a sequence of deductive logical statements, or syllogism, is that the conclusion is necessarily true if the premises of the argument are true. Truth must exist, therefore, before it is discovered by the human mind. This fact poses yet another question. In what does truth have its existence or in what does it subsist? As mentioned above, according to Dewy and Bentley in "Knowing and the Known", knowledge, per se, has no existence apart from mind. Therefore, truth has its only existence in a knowing mind. This conclusion has profound consequences when applied to the necessary pre-existence of infinite and eternal consilient truth. The misapprehension among some mathematicians and philosophers that new mathematics is invented derives from the impression of a creative event when new ideas arise in the mind of the mathematician by spontaneous insight. This phenomenon is discussed at length, and its relationship to inductive reasoning is considered. While deductive reasoning argues from a general truth to demonstrate a specific instance of that truth, inductive reasoning argues from a specific instance of truth to a general truth. The general truth is not immediately obvious but must be hypothesized, or inferred, which leads to the impression of discovery or invention.

Chapter 8 reviews the main conclusions of the preceding chapters as a prelude to consideration of the essential ontological question concerning the existence of God. The chapter begins with "The Metaphysical Poem of Parmenides", which in its own right provides an intriguing basis for belief in a necessary Being that is the source of all being. The main philosophical arguments for the existence of God are briefly presented, beginning with St. Augustine's original *Argument from Truth*, which is then followed in historical order by Boethius' *Argument for the Necessity of a Supreme Good*, and detailed explanation of St. Anselm's *Ontological Argument*.

The five arguments of St. Thomas are then mentioned with particular attention given to *The Argument of the First Cause* and *The Argument of Contingency,* which together lead to the existence of a Necessary Being that is the Self-Sufficient First Cause all that exists. St. Thomas's *Argument of the First Cause* and *Argument of Contingency* are re-evaluated in the context of the existence of an eternal multiverse, in which case we must ask whether such an entity could be the Necessary Being that is the sufficient cause of itself. Chapter 8 co-author, Arthur Kyriazis, and I then show why the interdependent collection of pocket or bubble universes that comprise the eternal multiverse cannot be the sufficient cause of itself. We then show why the parallel argument that atheists make against the self-sufficiency of God as the Necessary Being is answered by the *Modified Argument from Truth*, which we describe in detail. We show the essential role played by the existence of infinite and eternal Consilient Truth, and the epistemology of Dewy and Bentley which posits that knowledge, per se, has no existence of its own but rather must exist in *mind knowing truth.* This argument leads to the existence of an eternal mind that knows eternal Consilient Truth.

References

Burnet J (1920) Parmenides on nature. Early Greek philosophy, 3rd edn. A & C Black, London, pp 117–121

Conselice CJ, Wilkinson A, Duncan K, Mortlock A (2016) The evolution of galaxy number density at $z < 8$ and its implications. Astrophys J 830(2):83

Holt J (2012) Why does the world exist? Liveright Publishing Corporation, New York

Judson HF (1996) The eighth day of creation. Cold Spring Harbor Laboratory Press, Cold Spring Harbor, p 589

Premack DG, Woodruff G (1978) Does the chimpanzee have a theory of mind? Behav Brain Sci 1:515–526

Quine WVO (1948) On what there is. Review of metaphysics. Reprinted in: From a Logical Point of View. (1953) Harvard University Press, Cambridge, MA

Tajadura-Jimenez A, Banakou D, Bianchi-Berthouze N, Slater M (2017) Embodiment in a child-like talking virtual body influences object size perception, self-identification, and subsequent real speaking. Nat Sci Rep 7:9637. https://doi.org/10.1038/s41598-017-09497-3

Vaillant GE (1993) The wisdom of the ego. Harvard University Press, Cambridge, MA

Chapter 2
The Polarization and Reconciliation of Science and Religion

Richard J. Di Rocco

> *I find it as difficult to understand a scientist who does not acknowledge the presence of a superior rationality behind the existence of the universe as it is to comprehend a theologian who would deny the advances of science.*
>
> Werner von Braun (2007)

> *My own view is that, while science and religion may seem different, they have many similarities, and should interact and enlighten each other.*
>
> Charles Townes (2005)

> *Science without religion is lame. Religion without science is blind.*
>
> Albert Einstein (1941)

Abstract Science and religion seek an explanation for the basis of reality, but neither has the ability to objectively decide the ultimate question concerning God's existence. Theism and atheism are both based on faith; and, in the absence of convincing arguments, proponents of each have polarized in the heat of debate. The essential ontological question about God has been overshadowed by irrelevant arguments. The literal truth of the Bible, as well as the effectiveness of science as a method of understanding natural phenomena, have become proxy issues that are completely irrelevant to the existence of God. Consilience, which posits the unitary nature and coherence of all knowledge (Knowledge is *known* truth), assures us that the reconciliation of science and religion is possible insofar as practitioners of each

R. J. Di Rocco (✉)
Psychology Department, University of Pennsylvania, Philadelphia, PA, USA

Psychology Department, St. Joseph's University, Philadelphia, PA, USA
e-mail: richdi@upenn.edu

© Springer Nature Switzerland AG 2018
R. J. Di Rocco, *Consilience, Truth and the Mind of God*,
https://doi.org/10.1007/978-3-030-01869-6_2

discover the truth in varying degrees. A key insight provided by John Dewey's and Arthur Bentley's theory of knowledge, as described in "Knowing and the Known", is that the apprehension of truth by finite mind is always limited and imperfect. It is reasonable to assume, therefore, that as scientists, philosophers and theologians make progress incrementally in the understanding of ultimate truth, the statements they each make should increasingly demonstrate elements of correspondence. Until then it is clear that agnosticism offers an honest starting point in consideration of the question whether the universe the sufficient cause of itself, or whether it bears an imprint of a transcendental cause from beyond its spatial and temporal dimensions.

Keywords Atheism · Theism · Creationism · Intelligent design · Consilience · Epistemology

Science and Atheism

Before embarking on a discussion of the polarization between the proponents of science and religion, it is important to note that while many scientists consider themselves to be atheists this is by no means true of all scientists. Indeed, many famous scientists and mathematicians are known to have believed in the existence a supreme being or intelligence, even as others have denied it. Sir Francis Bacon, who is known as one of the founders of the rational empirical approach to scientific inquiry, stated in "Of Atheism" (Bacon 1597):

> It is true, that a little philosophy inclineth man's mind to atheism, but depth in philosophy bringeth men's minds about to religion; for while the mind of man looketh upon second causes scattered, it may sometimes rest in them, and go no further; but when it beholdeth the chain of them confederate, and linked together, it must needs fly to Providence and Deity.

Bacon anticipates the idea that a vital and persuasive clue to attaining proof of God's existence is found in considering the notion of consilient truth. Consilience refers to the coherent and unitary nature of all knowledge, or truth, the chain of causes "confederate, and linked together". (See the discussion of consilience below). Sir Isaac Newton, Blaise Pascal, Galileo Galilei, and Max Planck, also counted themselves among the ranks of believers. Planck held the opinion that both science and religion wage a "tireless battle against skepticism and dogmatism,[1] against unbelief and superstition." Plank held further that the goal of both science and religion was to attain "toward God" (Planck 1937).

---

[1]Traditional approaches to religion involve a substantial degree of dogmatism, however. It would be hard to defend Planck's position on a lack of dogmatism in religion. Most scientists also would deny that the goal of science is to "attain toward God", but if God is real and is the ultimate truth then Planck is correct in this assertion.

Albert Einstein characterized himself as being disdainful of the notion of the personal God who is portrayed in scripture. For example, in a letter to philosopher Eric Gutkind dated January 3, 1954, Einstein wrote:

> The word god is for me nothing more than the expression and product of human weaknesses, the Bible a collection of honorable, but still primitive legends which are nevertheless pretty childish. No interpretation no matter how subtle can (for me) change this. For me the Jewish religion like all other religions is an incarnation of the most childish superstition.

In a letter (Einstein 1954) to atheist Joseph Dispentiere Einstein wrote:

> It was, of course, a lie what you read about my religious convictions, a lie which is being systematically repeated. I do not believe in a personal God and I have never denied this but have expressed it clearly. If something is in me which can be called religious then it is the unbounded admiration for the structure of the world so far as our science can reveal it.

On the other hand, Einstein was not an atheist as indicated in the following cable reply to Rabbi Herbert S. Goldstein of the Institutional Synagogue in New York, who questioned Einstein, "Do you believe in God?" (Einstein 1929):

> *I believe in Spinoza's God who reveals himself in the orderly harmony of what exists*, not in a God who concerns himself with fates and actions of human beings. [Emphasis Added].

Purportedly, Einstein flatly denied being an atheist in an interview (Viereck 1929):

> *I'm not an atheist* and I don't think I can call myself a pantheist. We are in the position of a little child entering a huge library filled with books in many languages. The child knows someone must have written those books. It does not know how. It does not understand the languages in which they are written. The child dimly suspects a mysterious order in the arrangements of the books, but doesn't know what it is. That, it seems to me, is the attitude of even the most intelligent human being toward God. [Emphasis Added].

On the basis of this evidence, we can conclude that Einstein was not an atheist ("I'm not an atheist") and that he did believe in an intelligence that is the power responsible for the existence of, and the order in, the Universe ("Spinoza's God who reveals himself in the orderly harmony...").

The Limitations of Science and Religion for Proving the Existence of God

Modern physics and cosmology attempt to explain the fundamental nature of reality and the cause of the universe's origin, but can science succeed in studying such a cause if it lies outside of space and time? How can rational empirical science study a transcendent, extra-universal, cause of the universe when the methods of science must be practiced within it?[2]. This difficulty provides an illogical basis for the atheism of some scientists: *we cannot prove the existence of God by scientific methods so it is not a relevant question or, God must not exist.* The premise of the

---

[2]This difficulty will be addressed further in Chap. 3.

statement is likely true, but neither conclusion follows. The inability of science to effectively deal with the question of God's existence is a determinant of neither the question's importance, nor the answer to the question. Moreover, it is also true that the existence of God cannot be disproved by scientific methods. One might say as well, therefore, *we cannot scientifically disprove the existence of God,*[3] *so He must exist.* Here again the premise is true, but the conclusion does not follow. Since the existence or non-existence of God cannot be definitively decided by rational empirical methods, atheists who cloak anti-theist arguments in the guise of science are in danger of displaying a degree of hypocrisy and arrogance in their insistence that they know what they cannot prove. The conviction of the atheist appears, therefore, to rely on faith as much as the conviction of believers. One may say at least of believers that they know and freely admit that their belief is based on faith.

Religion, likewise attempts to provide an explanation for the origin of the universe and the basis of reality. Religion proclaims faith in a Supreme Being who is the cause of the universe, a Creator who encompasses the paradox of being both immanent and transcendent in relation to what has been created.[4] Many find comfort in such beliefs. Unfortunately, some believers go beyond comfort in their belief in the existence of God as justification for never having to wonder about anything again. Must belief in God suspend curiosity? This also is arrogance, and worse.[5] As with the faith of atheists, the faith of believers cannot be proven definitively by verifiable empirical methods. What then can be said with confidence about the theist-atheist debate?

False Arguments and the Straw-Man in the Polarized Debate

Before proceeding to examine the polarized debate between proponents of science and religion we should note what Princeton mathematician and physicist, Freeman Dyson, says about how these two radically different approaches to understanding reality can and do co-exist harmoniously in the minds of many people (Dyson 2000):

> Science and religion are two windows that people look through, trying to understand the big universe outside, trying to understand why we are here. The two windows give different views, but they look out at the same universe. Both views are one-sided, neither is complete. Both leave out essential features of the real world. And both are worthy of respect.
> Trouble arises when either science or religion claims universal jurisdiction, when either religious or scientific dogma claims to be infallible. Religious creationists and scientific materialists are equally dogmatic and insensitive. By their arrogance they bring both science

---

[3]In regard to the statement that, *"we cannot scientifically disprove the existence of God"*, it is important to note that it is notoriously difficult to prove the negation of a proposition. Moreover, rational empirical methods do not reach beyond the bounds of space-time. Science is necessarily practiced within the universe. The best science can do is look for an imprint of the transcendent on our universe, but again, failing to find such an imprint does not disprove the existence of God.

[4]Religion posits that God is immanent in the sense of being "intimately connected" to the universe, but simultaneously transcendent, by existing apart from it.

[5]It is willful ignorance!

and religion into disrepute. The media exaggerate their numbers and importance. The media rarely mention the fact that the great majority of religious people belong to moderate denominations that treat science with respect, or the fact that the great majority of scientists treat religion with respect so long as religion does not claim jurisdiction over scientific questions.

Unfortunately, the harmony of which Dyson speaks is not universal. Many theists are threatened by science, because they believe that science has the denial of God as a fundamental objective. This belief is encouraged by the atheistic writings of some scientists and philosophers. Notable among these are Bertrand Russell in "Why I am Not a Christian and other Essays on Religion" (Russell 1957); and more recently, Richard Dawkins in "The God Delusion" (Dawkins 2008); and Victor J, Stenger in "God: The Failed Hypothesis" (Stenger 2007). Theists should not be threatened by scientific explanations of natural history, however. The reduction of uncertainty that science provides does not inherently make God irrelevant, and certainly does not provide a proof of the nonexistence of God. How could it? Why would believers accept the premise that human understanding of reality is inconsistent with the existence of God? That is, why should believers allow themselves to be drawn into the argument that human understanding of the natural world precludes the existence of God? Scientific explanations of phenomena that humans seek to understand in no way calls God's existence into question, even though some scientists imply falsely that it does. The idea that God's existence requires that the workings of the universe are necessarily inscrutable is false.

Because they see the scientific narrative and program as a threat, many believers attempt to provide a reconstruction of the scientific view of creation and biology in support of their own interpretation and agenda. They are attempting to create a pseudo-science that passes for true science. They wish to present what appear to be rational arguments in favor of their point of view. In the case of many advocates of *creationism*, often presented in the guise of *intelligent design*, unfounded and un-testable conjectures are advanced as if they represent thoroughly tested consensus scientific opinion. This is extremely dangerous because it uses false arguments or premises to attack the findings and conclusions of rational empirical science, a method that has been shown time and again to overthrow erroneous hypotheses and reveal what appear to be true, albeit imperfect, insights into the workings of the natural world. Science is able to achieve this because it is based not only on opinion and hypothesis, but also on experimentation to validate or refute those hypotheses. Validation by other scientists through replication of results is required before a particular hypothesis is elevated to the status of *theory*.[6] The assertions of a theorist

---

[6]Science recognizes a hierarchy of the likely truth of statements about reality. *Hypothesis*, or *conjecture*, refer to an untested potential explanation of an observed phenomenon. At this stage, the proposed *explanation* is an inference that arises as a spontaneous insight that originates in the cognitive unconscious mind (see Chap. 7). Once experiment confirms the likely truth of the hypothesis it may be referred to as a *theory*, which is regarded as a reasonable explanation of a phenomenon, but subject to further validation by scientific replication and also subject to revision pending those future experimental results. Once an experimental result is validated by many different scientists and widely accepted as truth, the *theory* may be "elevated" to the level of a scientific law, as in Newton's *Law of Universal Gravitation*.

must stand up to empirical testing. Personal convictions, even well-founded hypotheses, cannot be presented as scientifically validated truth unless they have been repeatedly scrutinized and confirmed by scientific methods.

Opposed to the theists are atheists, some of whom also misuse science and fallacious arguments in an effort to prove that God does not exist. In this manner, they fuel the false impression of believers that science is necessarily opposed to, or contradicts, the existence of God. Such atheists, many of them scientists, foster the specious argument that somehow God becomes unnecessary because science is so successful in explaining natural phenomena. The fact that science can explain many natural phenomena simply demonstrates that humans are capable of using rational empirical methods to develop understanding of reality. It says nothing at all about whether a divine being or higher power exists and whether such a being created the universe. Moreover, it is by no means clear that the utility of science extends to the study and understanding of questions related to the existence of God, as already mentioned. For this reason, the scientist-atheist argument against the existence of God relies predominantly upon a case based on the utility of science as a method of inquiry that has validly refuted many dogmatic and false arguments advanced by believers over the centuries in an effort to establish the literal truth of the Bible. Refutation of such a *straw-man* by scientific methods is irrelevant, however, to the essential ontological question concerning the existence of God, which should be considered on its own merits independently of questions concerning the literal truth of the Bible, or the effectiveness of science as a method to discover new understandings of nature.

Consilience: The Unitary Nature of Knowledge

It is clear that misunderstanding dominates the dialogue between the practitioners of religion and science. From the foregoing, it is also clear that this misunderstanding must be attributed in large measure to the natural tendency of those who hold opposing opinions to state their views and those of their opponents in increasingly extreme form in the heat of debate. While this polarizes the discussion, the divergence of the narratives advanced by proponents of science and religion is not inevitable, nor does it necessarily represent a permanent and insurmountable divide. This idea may seem surprising owing to the clear disparity in the methods and tenets of these two approaches to understanding the basis of reality. On the other hand, while they have vastly different methods, both science and religion have common interest in understanding the ultimate meaning of reality, its origin and cause. With this in mind, it is possible to entertain the notion that ultimately both science and religion will provide an understanding of the same truth. Quoting William Whewell (Whewell 1840) Harvard socio-biologist E. O. Wilson defines consilience as (Wilson 1998):

> Literally a 'jumping together' of knowledge by the linking of facts and fact-based theory across disciplines to create a common groundwork of explanation. . . . .The Consilience of Inductions takes place when an Induction, obtained from one class of facts, coincides with an Induction obtained from another different class. Thus, Consilience is a test of the truth of the Theory in which it occurs.

Wilson builds a case for the idea that all knowledge is unitary and internally consistent by illustrating examples such as: the unification of genetics with Darwinian evolution; and the application of statistics to thermodynamics to produce statistical mechanics. The unification of general relativity and quantum mechanics is believed to exist and is sought by string theorists and others, but has yet to be proven. Knowledge refers to that which is believed to be true on the basis of some empirical evidence or logical proof. Consilience, therefore, is based ultimately upon the unity of truth. This follows from a definition of truth as *that which can be known without contradiction.*[7] It is already well established, for example, that *The Truth*[8] consists of statements that are internally consistent because true statements are never contradictory.[9] The contribution of Wilson's book and its development of the concept of consilience is that it points to and establishes the unitary nature and coherence of the findings of diverse areas of human inquiry, even when the body of findings from these diverse endeavors appears at first to consist of unrelated elements. Gerald Schroeder provides an interesting, and extremely inclusive, example of this convergent understanding by offering a scientific explanation of the creation story of Genesis (Schroeder 1998). Schroeder's thesis involves a necessary difference between the human perspective of space-time and God's. According to Schroeder, creation cannot be placed on a seven-day timeline and the universe is not literally five thousand years old from the human perspective within space-time. Instead Schroeder argues that the actual fourteen billion years since the origin of the universe, as observed from the human perspective, can be mapped onto a relativistic seven-day time-line that is valid from God's transcendental perspective. Schroeder's work shows that the consilience of knowledge provided by science and religion is possible.

Harvard Paleontologist, Stephen Jay Gould, was amenable to the compatibility of science and religion, but he referred to them as professing "non-overlapping magisteria"[10] (Gould 1997). We should note, however, that not only do the present teachings of science and religion differ, but also that the canons of reasoning in these two diverse ways of thinking, knowing and believing presently differ radically. In the light of consilience, however, we must go further. For in that light, the ultimate convergence of science and religion is inevitable to the extent that scientists, philosophers and theologians make progress in pursuing the discovery of truth

---

[7]Caution is warranted in "the knowing of truth", however, because human apprehension of truth is always incomplete. This follows from the fact that what can be demonstrated by logical argument is only as valid as the premises of the argument, and what can be known on the basis of empirical inquiry is always subject to revision in light of new discoveries.

[8]*The Truth* refers to the totality of truth, i.e. *Consilient Truth.*

[9]In logic, the Law of Contradiction states that contradictory statements cannot both be true. This is related to the Law of the Excluded Middle, which holds that a statement is either true or false. That is, a statement is either true, or its denial is true.

[10]Magisterium (magisteria, pl.) refers to the teaching of the Roman Catholic Church. Gould uses it here with license to mean *teachings* in the inclusive sense.

with integrity and some degree of success. While the premise and implications of consilience for the future convergence of science and religion on the ultimate truth seems plausible, can it be said that anything resembling the total, coherent and absolute truth exists? We may even question whether *Consilient Truth*[11] is finite in the event that the *infinite eternal multiverse*[12] is shown to exist. Finally, if *Consilient Truth* exists we are bound to ask, with Plato and the other philosophers, *where does it reside, in what does it subsist and have its being.*

Plato proposed that perfect truth exists in an ethereal *World of Ideals* or *Perfect Forms.*[13] Plato postulated an ontological dualism in which there are two types of reality. The first is the reality of sensory experience, the sensible world, while the other is the intelligible world or the world of ideas. The intelligible world is the world of eternal, immutable *forms*, which sometimes is referred to equivalently as the world of *ideals* or *ideas.*[14] Thus, the geometry student has a notion of *triangle*, and can draw a triangle in the sensible world, but the true and perfect archetype of *triangle* is understood to actually exist in a separate realm or reality, the *World of Forms, Ideals, or Ideas.* The perfection of the idea of *triangle* was described by Euclid in his plane geometry, but whenever a triangle is drawn it falls short of the perfect ideal because it is impossible to draw a line with length but no width.[15] The *Ideals* of Plato are abstractions, and everyone knows that *abstractions* and *ideas* are formed by mind and by nothing else. Concerning the Form of Beauty, however, Plato writes in the *Symposium*: "It is not anywhere in another thing, as in an animal, or in earth, or in heaven, or in anything else, but itself by itself with itself." An *abstraction*, does not exist *itself by itself with itself,* but is created by a cognitive act of mind. Despite the fact that what we call knowledge is widely understood to exist in human thought, knowledge is also commonly referred to as having an existence outside of mind. Consider for example the statement, *mankind has generated a large body of knowledge since the advent of empirical science.* This usage of the word is hard to avoid, and requires that we ask where this knowledge resides. Plato would

---

[11]*Consilient Truth* is used in this book to refer to the totality of what can be known about logic and reality without contradiction. *Consilient Truth* is therefore coherent.

[12]The theory of the *eternal multiverse* is a proposal in the field of cosmology which states that the universe we can observe is merely one of an infinite number of universes each one of which is spawned by a preceding one. The eternal, self-replicating multiverse is necessarily infinite as well. The truth that can be known about the multiverse also must be infinite.

[13]*World of Ideals or Perfect Forms* is Plato's conception of a reality different from the corrupted one in which humans live and within which we attempt to discover truth. See the text for further explanation.

[14]The words "ideas" and "intelligible" imply an act of mind, not a realm that exists in and of itself as the domain in which Truth exits eternally.

[15]In his 13 books on plane geometry, "The Elements", Euclid defined a point as "that which has no part", i.e. it lacks dimension but has location in the plane. It is fascinating that Euclid defined a point in a manner that is consistent with the rigorous understanding of the *real number line* that was developed approximately 2000 years later. Euclid defined the geometric concept of a line as "breadthless length", i.e. it has length but no width. Therefore, any geometric figure that we draw is imperfect, but the idea of that figure exists in the World of Perfect Forms, Ideals, Ideas, etc.

say that the world of forms contains the perfection of all knowledge and truth, but knowledge is "known truth" and necessarily requires the mind that knows it. Therefore, knowledge necessarily subsists only in mind in the cognitive act of "knowing". To speak of "knowledge" as having an existence of its own, apart from the mind of a knower, is a concept that is rejected in "Knowing and the Known" by John Dewey and Arthur Bentley, as discussed below and in Chap. 7.

Insight from Epistemology

In their landmark treatise on epistemology, "Knowing and the Known" (Dewey and Bentley 1949), Dewey and Bentley argued for two premises of great importance to the theory of knowledge. As explained above, the first is that knowledge has no existence independent of knowing mind. The second premise that Dewey and Bentley argued is that the discernment of truth through various processes of human inquiry proceeds incrementally, and that the state of perfect understanding is always approximated but never achieved. In other words, all methods of human inquiry necessarily yield incomplete results. The history of science certainly reveals this to be true. Theories arise and appear to explain phenomena of interest only to be overturned by newer and more complete theories that provide a deeper understanding of the phenomena, often after the problem situation itself is recast and understood in new terms that provide a basis for further investigation along new lines of inquiry. Perhaps the most recognized example of a theory to be so displaced is the mechanics and associated theory of gravity advanced by Sir Isaac Newton, which was superseded by Albert Einstein's theories of Special and General Relativity. In the spirit of Dewey's and Bentley's epistemology we may argue that, while all knowledge and comprehensive truth is unitary as required by consilience, it can never be understood or known in its entirety by finite mortal mind. Yet according to the first premise, Knowledge and Truth exist exclusively in mind! The concept of consilient truth therefore leads to a contradiction between Dewey's and Bentley's requirement that knowledge, or known truth, must exist exclusively in mind and the necessary incompleteness and incremental nature of human apprehension of truth. These issues will be explored in greater detail in Chap. 7.

Program for Reconciliation

Such considerations lead to the realization that both science and religion, irrespective of their relative merits, necessarily provide only partial or incomplete understanding of ultimate truth. One can also expect that if practitioners of science and religion make progress in discovering and understanding truth through diverse methods, the insights that they each discern will begin to approximate the ultimate comprehensive truth ever more closely, and therefore their respective statements will become increasingly similar. There is only one Consilient Truth! Religion based on revelation and faith, and science based on rational empiricism, have methodologies and canons of reasoning that are so vastly different that it is not surprising that practitioners of each often disagree. Yet even now we see that science and religion have common interest in questions related to the origin of the universe, the origin of life, the origin of human life and sapient mind, the ultimate fate of the universe, and so on. If consilience is valid, the vastly different understanding of reality provided by religion and science is evidence only of the incompleteness of the discernment that they each offer at the present time.[16] Regrettably, an excess of emotion, intolerance, and irrationality have fueled antagonism and impeded progress along lines of common interest.

If the Universe was created by a Supreme Intelligence, as people of faith believe, then that same Intelligence also established the natural laws and physical constants at the moment of creation. Physicist John Barrow, among many others, has noted that many physical constants have precise values that are necessary for stars to form and for life and mind to evolve from inanimate matter that was present in the very early universe (Barrow 1994). These fundamental laws and physical parameters determine the way that the universe unfolds, how it evolves. If the universe exists by virtue of the "intelligent design"[17] of a divine creator, design would have been operational at the inception of the Big Bang[18] when all natural law was established, or became operational in our universe. In any event, it is generally accepted that from its beginning the universe has evolved according to natural law. This involves the processes and phenomena that empirical science and mathematics have revealed. Perhaps most amazing among these is the coherence and relevance of mathematics as a language that not only models and explains phenomena in the world, but also predicts new knowledge based on mathematical operations on the elements of those models (King 2009). Nobel Laureate in physics, Eugene Wigner, commented in regard to this (Wigner 1960):

---

[16]Scientists will readily admit that what has been revealed as truth through the empirical enterprise is incomplete, and certainly believers who hold to a dogmatic theological understanding of reality must admit that they do not know everything that can be known about God.

[17]If created, the universe was designed on the basis of some intelligence. Thus, there is a valid form of *Intelligent Design,* not corrupted by false agendas or false statements of creationists who seek to establish a false belief system in support of what they claim is the true one, i.e. the literal interpretation of the Bible.

[18]The Big Bang" is the event described in modern cosmology as the beginning of what Einstein called Space-Time, the four-dimensional universe in which we live. It is believed that all matter and energy as well as the laws of physics were created and appeared at the inception of the Big Bang. At the moment of creation, the universe, all of its matter and energy were condensed into an infinitely small point which has been expanding ever since. The progression of Natural History involves the expansion of three-dimensional space and the progression of time. No one has yet explained how everything appeared at the inception of the Big Bang or from where it came.

> The enormous usefulness of mathematics in the natural sciences is something bordering on the mysterious and there is no rational explanation for it. It is not at all natural that the 'laws of nature' exist; much less that man is able to discover them. The miracle of the appropriateness of the language of mathematics for the formulation of the laws of physics is a wonderful gift which we neither understand nor deserve.

Subsequent empirical validation of predictions derived from mathematics provides a powerful demonstration of the coherence of mathematics, as well as the precise correspondence between mathematics and the real world. Other key factors that govern the progression of natural history are the fundamental forces and laws of physics[19]; the randomness of quantum mechanics; random genetic mutations and natural selection of adaptations that confer advantages on the organisms that possess those adaptations.

Many believers cannot accept the random aspect of mutation as a tool or mechanism of evolution because they see it as contradicting *design*. They believe that God doesn't need *mindless randomness* to achieve His objectives. Even Albert Einstein is reputed to have been disturbed by the probabilistic nature of quantum phenomena and is said to have remarked, "God does not play dice with the universe" (Isaacson 2007a, b). Upon hearing Einstein's comment Niels Bohr, one of the founders of quantum mechanics, is reputed to have quipped in reply, "Einstein, stop telling God what to do" (Ibid). Clearly, many individuals of all stripes, especially believers, are uncomfortable with the random aspect of many natural phenomena, but in the divine creation scenario, God establishes the laws of mathematics and probability too. If sapient beings are to have free will, the unfolding of the Universe must not be absolutely deterministic. There must be some element of unpredictability or uncertainty at a deep fundamental level of reality. This may be provided by the probabilistic nature of quantum mechanics at the atomic and sub-atomic scale. *Uncertainty*, and the related concepts of *information* and *entropy*, as well as some of the basic tenets of quantum theory will be considered in the next chapter. We will see that each of these concepts of physics has important implications for understanding the emergence and evolution of life and mind that will be considered in Chaps. 5 and 6.

An Honest Starting Point – Agnosticism and Radical Amazement

We may never fully prove or disprove the existence of God, or understand the mind of God, by the methods of rational empirical science, if indeed God is real. Nor is it likely that dogmatic faith-based religion will provide a basis for *objective* belief in

---

[19]The laws of physics are expressed most efficiently not in words, but in the language of mathematics. What is the most essential expression of these laws? Is there a fundamental, irreducible, core of physical law expressed in mathematics from which all truth may be derived? Does consilient truth have a deep structure that provides the basis upon which all of reality exists?

God.[20] It would seem, therefore, that the only initial logical position in regard to the existence of God is *agnosticism*.[21] That is to say, honest inquiry into this subject must at least begin with an open mind. A mind without prejudice regarding the question of God's existence will approach the issue with curiosity and honesty. This is a good point to ask yourself whether you are open to the possibility that God is real. If you answer no because you hold a confirmed atheistic conviction, then I hope you will at least be filled with a legitimate sense of awe and radical amazement, when contemplating the incomprehensible vastness and beauty of the universe. Abraham Joshua Heschel wrote eloquently of faith having its origin in this sense of radical amazement in the contemplation of the universe (Heschel 1995). Surely there is a basis that we all can find for the intellectual honesty required to admit that the universe constitutes an awesome breath-taking reality that requires an explanation, whatever that may be.

A Discernible Imprint of Creation?

Would an imprint of a transcendental causative agent on the cosmos provide evidence of a confluence of science and religion? In his "Letter to the Romans", the Apostle Paul makes the point that the existence of the universe *necessarily* implies the existence of God when he says,

> For what can be known about God is plain to them [the pagans], because God has shown it to them. Ever since the creation of the world his invisible nature, namely, his eternal power and deity, has been clearly perceived in the things that have been made (Rom 1:19–20).

Paul's thesis implies that there is a fundamental and intimate relationship between God and the universe. According to Paul, God's "deity and power" are manifested in His creative activity, the expression of His Divine thought, because it can be discerned by contemplation of "the things that have been made". Was Paul correct in his belief that the universe bears an imprint of a transcendental causative agent? To help answer this question, we must delve into what is known about the universe. The next two chapters are devoted to an examination of some of the key findings of physics and cosmology. There, we will see that scientists are indeed looking for such an imprint; and that some believe it will soon be found. We will then begin an exploration of biology and psychology to consider the origin of life and the crowning achievement of biological evolution in the emergence of intelligent mind. The sapient mind possessed by modern humans is capable of

---

[20]While neither rational empirical science nor dogmatic faith based religion is likely ever to prove or disprove the existence of God, philosophy *is* competent to decide this ultimate ontological question, as we will see in Chap. 8.

[21]*Agnosticism* is the philosophical perspective that the existence of God is indeterminate or, in the extreme, unknowable. The term is used here in the former sense to describe the position of an individual with an open mind about whether God is real or not.

exploring questions about the origin of the universe, life, mind itself, and how that mind can be used to evaluate the ultimate ontological question concerning the existence of God.

References

Bacon F (1597) Of Atheism. In: Meditationes sacrae and human philosophy. Kessinger Publishing, LLC, Facsimile edn, February 1, 1996, p 70

Barrow JD (1994) The origin of the universe. Basic Books, New York, pp 124–125

Dawkins R (2008) The god delusion. Houghton Mifflin, New York

Dewey J, Bentley A (1949) Knowing and the known. The Beacon Press, Boston

Dyson F (2000) Templeton prize acceptance speech

Einstein A (1929) Einstein Believes in Spinoza's God. New York Times, April 25, 1929 p60 col 4

Einstein A (1941) Science and religion. In: Science philosophy and religion, a symposium. The conference on science. Philosophy and Religion in Their Relation to the Democratic Way of Life, Inc, New York

Einstein A (1954) In: Dukas H, Hoffman B (eds) (1981) Albert Einstein, The human side. Princeton University Press, Princeton, p 43

Gould SJ (1997) Nonoverlapping Magisteria. Nat Hist 106(March):16–22 and 60–62

Heschel AJ (1995) God in search of man, 17th edn. The Noonday Press/Farrar, Straus and Giroux, New York

Isaacson W (2007a) Einstein: his life and universe. Simon and Schuster Paperbacks, New York, p 84

Isaacson W (2007b) Einstein: his life and universe. Simon and Schuster Paperbacks, New York, p 326

King J (2009) Mathematics in 10 easy lessons. Prometheus Books, Amherst/New York, pp 19–26

Planck M (1937) Religion and natural science. Paper presented in the Baltics

Russell B (1957) Why i am not a christian. Simon and Schuster, New York

Schroeder G (1998) The science of god: the convergence of scientific and biblical wisdom. Broadway Books, New York

Stenger SJ (2007) God, the failed hypothesis. Prometheus Books, Amherst

Townes C (2005) Speech upon accepting the Templeton Prize

Viereck GS (1929) What life means to Einstein. Saturday Evening Post, p 17. Quoted by D. Brian, Einstein: a life, p 186

von Braun W (2007) In: Irene E. Powell-Willhite (ed) The voice of Dr. Werner von Braun: an anthology, p 89

Whewell W (1840) The philosophy of inductive sciences, founded upon their history, in two volumes, London

Wigner E (1960) The unreasonable effectiveness of mathematics in the natural sciences. In: Communications in pure and applied mathematics, vol. 13, No. I. Wiley, New York

Wilson EO (1998) Consilience: The unity of knowledge. Vintage Books, New York, pp 8–9

Chapter 3
Physics

Richard J. Di Rocco

> *I want to know how God created this world. . . . I want to know God's thoughts.*
>
> Albert Einstein (2000)
>
> *As the heavens are higher than the earth, so are my ways higher than your ways and my thoughts than your thoughts.*
>
> Isaiah 55:8–9
>
> *Every picture tells a story.*
>
> From the song by Rod Stewart

Abstract A review is presented of the highlights of the natural history of the universe from the inception of space-time known as the Big Bang. The implications of this knowledge are presented in subsequent chapters that discuss cosmology, and the origin of life and sapient mind from inanimate matter. The present chapter begins an overview of classical physics with brief mention of the vital contributions of Galileo's early studies of motion, which laid the groundwork for Newton's far-reaching mechanics and co-discovery of Calculus. The grand achievements of Eighteenth and Nineteenth Century physics, thermodynamics and statistical mechanics, are then discussed with particular attention paid to the related concepts of entropy and information that are essential to consideration of the origin and evolution of life and mind in later chapters. After mention of James Clerk Maxwell's theory of electromagnetism, consideration is given to some of the key elements of quantum physics, in particular, the paradox presented by the simultaneous wave-particle duality of light and matter. The chapter concludes with a discussion of the phenomenon of

R. J. Di Rocco (✉)
Psychology Department, University of Pennsylvania, Philadelphia, PA, USA

Psychology Department, St. Joseph's University, Philadelphia, PA, USA
e-mail: richdi@upenn.edu

© Springer Nature Switzerland AG 2018
R. J. Di Rocco, *Consilience, Truth and the Mind of God*,
https://doi.org/10.1007/978-3-030-01869-6_3

quantum entanglement, which in its most extreme form claims that everything is connected and has a coherent unitary nature at the deepest level of reality.

Keywords Information · Entropy · Arrow of time · Quantum phase entanglement · Dual nature of light and matter · Photoelectric effect · Interpretation of quantum mechanics

Understanding Begins with Acceptance of Reality as We Find It

In considering the existential questions that perplex humanity, we can do neither more nor less than take reality exactly as we find it. We must take account of the universe and the laws of physics that govern it as expressed in the universal language of mathematics. All that we can know about reality with empirical validity derives from observation, hypothesis and experiment. At the moment of the universe's creation, space and time, all matter and energy, acquired the reality of existence in an infinitely condensed form or state which has been evolving according to physical law ever since. There is broad scientific consensus that approximately 13.7 billion years ago the universe, consisting of *space-time* together with all the matter and energy that exists, sprang into being in a cosmogenesis event known as the Big Bang. Space-Time is the four-dimensional reality, three of space and one of time, in which the universe evolves, and in which everything we can directly observe exists. The notion of space-time plays a major role in Einstein's theories of Special and General Relativity, but the idea of four-dimensional space-time was known and developed by notable mathematicians and physicists before him. The first manifestation of the Big Bang is referred to as a singularity, which is characterized by the Theory of General Relativity as a locus of matter and energy that approximates infinite density in an infinitely small volume. In fact, however, Einstein's equations for General Relativity break down and do not adequately describe the state of the universe in its first moment. For this a more comprehensive theory that is consistent with General Relativity and Quantum Mechanics is required. According to the current understanding of Quantum Mechanics, however, the initial volume of the universe cannot have a diameter smaller than the Planck length ($1.61619926 \times 10^{-35}$ meters). How such a massively dense and energetic entity first appears is still unknown. Black Holes, caused by the gravitational collapse of stars of sufficient mass, are also singularities, and subject to the same uncertainty regarding the actual size and density. In any case, it is clear that Space has been expanding since the Big Bang. A special aspect of this expansion, called *cosmic inflation*, will be examined in more detail in Chap. 4.

   The laws of physics, which consist of all the rules by which the universe unfolds, were operative at the first instant of the Big Bang and perhaps even before if our universe sprang into existence from one that already existed as postulated by the theory of the multiverse as explained in Chap. 4. The consensus view of modern physics and cosmology is that these laws supervene everywhere for all time to govern, proscribe and describe reality. Neither the existence of the laws of physics, nor the effectiveness of mathematics as a language that accurately expresses those laws has been explained. The ability of mathematics to discover new physical truths from symbolic operations on the objects of mathematics is likewise without explanation. So lacking explanation, and so remarkable is the effectiveness of mathematics in this regard that it was considered unreasonable by Nobel Prize winning physicist Eugene Wigner, as explained in Chap. 2.

Does the existence of a coherent structure of physical law that operates on a universal scale for all time, together with the existence of a coherent and logical mathematical language that expresses those laws and allows humanity to discover new ones, suggest an underlying unity or plan for the structure and workings of the universe? This is essentially the question of *design* and, rather than grapple with it now, it is more productive at this point to ask a related question.[1] To momentarily side-step the issue of design, the question may be reformulated as follows. Does the pervasive operation of physical law throughout space and time, together with the universal relevance and coherence of mathematics, suggest that everything is connected in some way in the deep structure of physical reality? That is, does reality have a unitary nature? Are all things connected, as all truths about those things likewise are connected? This is an important question. It asks if reality is as connected and coherent as the consilient truth that can be known about it. Many physicists believe that the question can be answered, and indeed has been answered substantially, in the affirmative on the basis of results obtained using the methods of rational empirical science. Quantum physics demonstrates that there is a deep fundamental connectedness among the constituents of reality. This connectedness is the property referred to in quantum theory as *quantum phase entanglement*, which is explored below. To understand its key points, it is necessary to review some basic aspects of classical physics and, in particular, quantum physics. This subject matter has been discussed at great length in a burgeoning literature written by the physicists themselves for the lay public over the last 150 years. One notable early example in the referenced time-frame is "The Theory of Heat", by James Clerk Maxwell which was written with non-physicists in mind (Maxwell JC, 1871). The full breadth of physics is obviously beyond the scope of this book, but the brief and selective overview presented here will facilitate understanding of what follows.

Classical Physics

What we now call classical physics advanced tremendously with the work of Galileo Galilei who lived during the latter part of the Sixteenth and early Seventeenth Centuries. Among other achievements such as the invention of the telescope, Galileo overturned Aristotle's false idea about motion requiring the constant application of a

---

[1]The question of design is a surrogate for the question concerning whether God exists. If indeed there is a Supreme Intelligence responsible for creation, then of course design is real and even the most avowed atheist would concede that such a being is God although not perhaps the anthropomorphized God that is conceived by many believers. If God does not exist neither does *design* as a deliberate plan in the mind of a Creator. In the absence of design, however, the laws of physics remain unexplained.

force. Sir Isaac Newton, who was born the year after the death of Galileo, built upon the work of his illustrious predecessor to advance a mathematical formulation of the laws of motion and gravity in "Philosophiae Naturalis Principia Mathematica",[2] often referred to simply as "The Principia". So valid and enduring in fact is Newtonian mechanics, which culminated in his Theory of Universal Gravitation, that it was used to put men on the moon almost 300 years after it was formulated. While his achievements in science and mathematics were extraordinary by any measure, Newton himself was said to have remarked, "If I have seen further than others, it is by standing upon the shoulders of giants."[3] In addition, Newton along with Gottfried Leibnitz was a co-discoverer of the calculus. Calculus is a branch of mathematics that has been essential for all subsequent work in physics, as well as many other branches of science.

Reversibility of Newtonian Mechanics and the Irreversibility of Time

Newton developed equations of motion that described how objects move in response to forces that act upon them. An example from the game of pool is illustrative of the general idea. When the cue ball is struck by a player's pool stick the force of the impact sets it in motion to move toward and strike the racked pool balls at the particular angle, and with the force, desired by the player. The force of the impact sets the racked pool balls in motion so that they each move around the pool table until friction with the table and air causes them to stop moving. During this process, some of the pool balls will strike the edge of the pool table and bounce off at an angle determined by the angle of the impact. In a similar manner, some of the pool balls may strike others and each will move away from the collision with a velocity (speed and direction) determined by the velocity of each pool ball at the moment of impact.

An interesting property of Newton's laws of motion is that they are equally valid for describing the motion in a video or film recording of a cue ball striking a single pool ball whether we play the video in forward or reverse motion. We begin our video viewing after the white cue ball is already set in motion toward a single stationary pool ball. We observe that the cue ball moves toward and strikes the pool ball, after which they both move off in directions determined by the direction of the cue ball at the moment of impact. In reverse motion of the video recording, the pool ball and the cue ball converge to a point of collision after which the pool ball stops moving and the cue ball continues to move[4] until the video ends. An observer of

---

[2]Translated as *Mathematical Principles of Natural Philosophy*. The book was published in 1687, when physics was called natural philosophy. The title may therefore be understood as *Mathematical Principles of Physics*.

[3]From a letter to Robert Hooke, 15 February, 1676.

[4]Recall that the forward motion video begins after the cue ball is set in motion so it is moving at the beginning of that video and at the end of the reverse video.

either scenario would not be able to tell whether the video recording was played in reverse or forward motion. This reflects the fact that Newtonian mechanics applies to and describes the motion of the cue ball and pool ball equally well in either case, and there is also no obvious indicator of the direction of time in either case. On the other hand, if a videographic recording were made starting when the cue ball approached and then struck a set of racked pool balls, an observer of forward and reverse versions of the recording could easily identify the forward and reverse directions of time in the two scenarios. This follows from the fact that we are accustomed to seeing objects disperse to become more disordered over time, but not to converge into a more ordered spatial structuring. The pool balls have an extremely low probability of moving toward each other, from various positions on the table, speeding up during the approach and then stopping abruptly at the exact moment the racked pattern is achieved and their collective momentum is transferred to the cue ball which then accelerates away from the point of contact with the racked pool balls! Newton's laws of motion describe either scenario equally well, but we know that there is a universal tendency for a system such as the pool balls to become more disordered in the forward motion of time. On the other hand, reverse viewing of the initial break of the racked pool balls by the cue ball, while adequately described by Newton's laws of motion could be instantly recognized as a reverse viewing of the actual events for reasons explained below.

Another example illustrates the irreversibility of time and the tendency toward increasing disorder of a system in time more dramatically. It also reveals something vital about the idea of information and its relationship to the direction of time. Consider the fall of the nursery rhyme character, Humpty Dumpty. The legendary nursery rhyme describes the irreversible existential fate of the unfortunate egg. There are two instructive aspects of this nursery rhyme: the catastrophic consequences that the fall has on the egg; and the inability of anyone to reassemble the egg. Watching a video recording of an egg falling from a table from beginning to end in forward motion makes perfect sense to the viewer. The reverse play of the recording, however, is obviously nonsensical because broken eggs that have fallen to the ground do not spontaneously reassemble themselves and then rise up in defiance of gravity to their former height on the table top. Every child understands this. Indeed, it is observations of events such as this that lead to the development of a child's notion of causality in the interactions of objects, and people, in the world as time advances forward. More important for our purposes than the fall of an egg and its result, however, is the apparent impossibility of reassembling the egg. That is, while we appreciate that eggs that have fallen do not spontaneously reassemble themselves, something else prevents their deliberate reassembly by some agent. That "something else" is the information needed to complete the desired reconstitution. In fact, considering all of the molecules in the dispersed yolk albumin, membranes and shell of the egg, an impossibly large amount of information would be required to reconstitute the egg by exactly reversing the molecular motions that occurred during its destruction. This example illustrates an undeniable truism: spontaneous events that occur for any system are associated with an increase in disorder and an increase in the amount of information that would be required to describe the condition of the system as it progresses to increasingly disordered states. This is the same amount of information that would be required to reverse the disordered condition to reestablish the original state of the system. It is highly unlikely that a broken egg could ever be reconstituted to its original state because the amount of information required to do so is extremely large and virtually impossible to ascertain and use.

These examples from the game of pool and the breaking of eggs provide good intuitive descriptions of the relationship between disorder, also known as entropy, information and their intrinsic connection to the forward direction of time. This forward direction of time has been referred to as the "arrow of time". Sir Arthur Eddington was the first to use this term (Eddington A S 1928). Eddington expresses the concept of the arrow of time using the terms, "random" and "randomness" to convey the sense of "disorder":

> Let us draw an arrow arbitrarily. If as we follow the arrow we find more and more of the random element in the state of the world, then the arrow is pointing towards the future; if the random element decreases the arrow points towards the past. That is the only distinction known to physics. This follows at once if our fundamental contention is admitted that the introduction of randomness is the only thing which cannot be undone. I shall use the phrase 'time's arrow' to express this one-way property of time which has no analogue in space.

We can summarize the foregoing by saying that when any system evolves over time the disorder, or entropy, of that system increases and the amount of information that would be required to describe the state of the system increases accordingly. This is the same amount of information that would be required to return the system to its original state.

We can see how the ideas concerning entropy, disorder and information emerged gradually and were formalized in the Nineteenth Century from the study of thermodynamics, which together with Maxwell's theory of electromagnetism was the crowning achievement of physics to that point in time. These ideas are presented with more formal detail in the next two sections. If you want to avoid the math, you can skip ahead to the section on quantum mechanics, which begins with the section headed "The Dual Nature of Light and Matter".

The First and Second Laws of Thermodynamics

The Eighteenth and Nineteenth Centuries brought an acceleration of progress in physics. The new field of thermodynamics provided an understanding of temperature and heat flow on a macroscopic or non-atomic scale. The First and Second Laws of Thermodynamics laid the groundwork for many of the technological innovations of the Industrial Revolution, such as the steam engine that converted heat energy to mechanical energy. The First Law of Thermodynamics states that energy can neither be created nor destroyed, but it can be converted from one form to another. This can be illustrated by considering the change in the internal energy content, U, of a steam engine as it burns fuel and does work. The burning of fuel increases the energy content of the boiler of a locomotive in the form of heat, Q. On the other hand, the movement of the locomotive's wheels that is caused by the pressure of steam heated by the boiler removes heat energy, in the form of mechanical energy, to perform work, W. The First Law of Thermodynamics states that the change[5] in the internal energy, ΔU, of a system such as a steam engine can be determined at any time by subtracting the change in the amount of work done by the system, ΔW, from the change in the amount of heat in the system, ΔQ. For the example of a locomotive, ΔQ is the amount of heat added to the system by the burning of fuel, and ΔW is the amount of work done by steam in moving the wheels. The First Law of Thermodynamics is captured in the following equation:

$$\Delta U = \Delta Q - \Delta W \qquad (3.1)$$

When the First Law of Thermodynamics was formulated, it was known that work is equal to force times the distance over which the force is applied:

$$W = F \times D \qquad (3.2)$$

Unlike work, however, the nature of heat was poorly understood until the latter half of the Nineteenth Century. The dominant theory had been that heat is a fluid called caloric. An early indication of the alternative, and correct, idea that heat is related to motion comes from the work of Count Rumford, born Benjamin Thompson, in 1753. Rumford had been working on the boring of cannon and was led to question the theory of caloric on the basis of his observation that an inexhaustible amount of heat could be produced by the continuous friction generated when boring the cannon barrel. This led him to conclude (Rumford B 1798),

It is hardly necessary to add, that anything which any insulated body, or system of bodies, can continue to furnish without limitation, cannot possibly be a material substance; and it appears to me to be extremely difficult, if not quite impossible, to form any distinct idea of anything capable of being excited and communicated in the manner the Heat was excited and communicated in these experiments, except it be motion.

Remarkably almost 200 years earlier in his treatise on inductive reasoning, "Novum Organum", Sir Francis Bacon reached the exact same conclusion based on his application of the inductive method to the question of heat (Bacon F 1620):

...the nature whose limit is heat appears to be motion. This is chiefly exhibited in flame, which is in constant motion, and in warm or boiling liquids, which are likewise in constant motion. ...What we have said with regard to motion must be thus understood, when taken as the genus of heat: it must not be thought that heat generates motion, or motion heat (though in some respects this be true), but that the very essence of heat, or the substantial self of heat, is motion and nothing else.

At first, he correctly observes that flame and boiling water are in constant motion to support his claim that, "the nature whose limit is heat appears to be motion". Then he seems to reach the opposite conclusion in the last paragraph where he says, "it must not be thought that heat generates motion, or motion heat (though in some

---

[5]Change is signified in scientific notation by the Greek letter delta, Δ.

respects this be true)"; but then goes on to state that, "the very essence of heat, or the substantial self of heat, is motion and nothing else." Bacon's understanding that the nature of heat is intimately connected to motion, while correct, seems to have been somewhat tentative. His inductive method clearly led him to the right inference about the relationship between motion and heat, but he was unable to bring the concept to complete fruition in the absence of an atomic theory of matter. "Novum Organum" is discussed further in Chap. 7.

The fog of ambiguity concerning the precise nature of heat was finally lifted when an atomic-scale mechanistic theory was formulated to explain what heat is and how it is transferred between bodies that have different temperatures. This theory was advanced in 1867 when James Clerk Maxwell postulated that molecules of a gas have a velocity-dependent energy of motion, called *kinetic energy*,[6] just as a macroscopic object like a falling apple does (Maxwell J C 1867). The kinetic theory of heat is explained clearly for the lay person in his famous book, "Theory of Heat" (Maxwell J C 1871). The vast number of atoms, even in a very small volume of gas for example, makes specification of the individual velocities and kinetic energies impossible for each of the atoms of the gas. So, Maxwell developed a statistical approach to describe the distribution of molecular velocity for a system of gas molecules in thermal equilibrium[7] at three different temperatures, as shown in the Fig. 3.1.

From this distribution of molecular velocities, Maxwell could calculate the average velocity of a closed system of gas molecules at uniform temperature throughout. Knowing this, Maxwell was able to determine the average kinetic energy of those same gas molecules since kinetic energy is:

$$K = \tfrac{1}{2}m\left(v_{avg}\right)^2 \qquad\qquad (3.3)$$

Where K is kinetic energy, m is the mass of a molecule of the gas and v_{avg} is the average molecular velocity. From such an approach, Maxwell developed an understanding of the macroscopic phenomena of thermodynamics in terms of a statistical treatment of the kinetic energies of molecules that is inherent to their motion. Maxwell explained that heat is nothing more than the collective mechanical effect of the kinetic energies of all the molecules in a gas, liquid, or solid that is being observed.[8] This new theory of heat came to be known as statistical mechanics or statistical thermodynamics. When a hot body is brought into contact with a cooler one, kinetic energy transfers from each body to the other as their respective molecules collide. Heat transfers from the hotter body to a cooler one because, during those collisions the *net* transfer of kinetic energy is from the hotter to the cooler one.

---

[6]Kinetic energy is the energy a mass possesses by virtue of its motion.

[7]Thermal equilibrium for an isolated system of gas molecules occurs when the temperature is uniform throughout the volume of gas.

[8]Heat therefore is expressed in units of energy, the Joule or J.

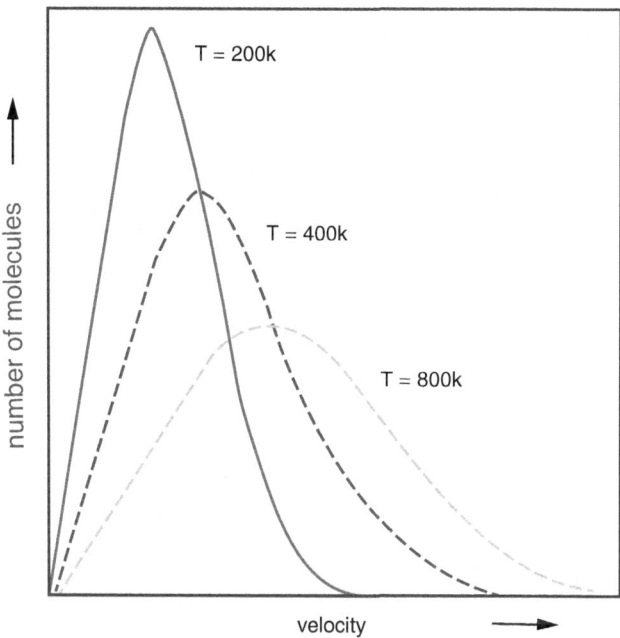

Fig. 3.1 The graph shows the distribution of molecular velocities for an enclosed system of gas molecules in thermal equilibrium at three different temperatures. Note that more molecules have higher velocities when the gas is hotter. Graph obtained from: "Quantum Physics, Thermodynamics, and Information." Image downloaded from Information Philosopher at http://www. informationphilosopher.com/quantum/physics/ and reprinted here under terms of the Creative Commons License found at: https://creativecommons.org/licenses/by/3.0/legalcode

The Second Law of Thermodynamics states that any natural or spontaneous process moves in the direction that causes the entropy of the system plus the environment to increase. Entropy was originally understood as a thermodynamic variable that described an entire system of molecules. Such variables as temperature and entropy of a body are therefore considered to be state variables of the system. In classical thermodynamics, entropy is defined in terms of heat and temperature. For each amount of heat (ΔQ) added to a volume of water, the entropy change (ΔS) is calculated according to:

$$\Delta S = \Delta Q / T \qquad (3.4)$$

where T is the temperature in degrees Kelvin at which the heat is added.[9] This is the macroscopic thermodynamic understanding of entropy, but in the hands of Ludwig Boltzmann, Maxwell's statistical mechanics would provide another far-reaching

---

[9]Entropy, S, therefore has the units of Joules per degree Kelvin or J/oK.

understanding of entropy (Boltzmann L 1877). Maxwell had derived the mathematical expression for the distribution of atomic velocities in a gas at thermal equilibrium, as shown in Fig. 3.1. In a volume of gas molecules that is removed from thermal equilibrium,[10] Boltzmann showed that the distribution of molecular velocities gradually approached that of Maxwell's distribution as a result of molecular collisions that bring the gas back into thermal equilibrium. In doing so, Boltzmann's theory provided a new understanding of entropy as molecular disorder which reaches a maximum value at thermal equilibrium. Boltzmann defined entropy as:

$$S = K_B \log_b(W), \tag{3.5}$$

where W is a measure of molecular disorder, expressed as the number of equally probable microstates[11] the atoms in a system can assume, and where K_B is a number known as Boltzmann's constant.[12,13] Implicit in this formulation of entropy is the understanding that the system at thermal equilibrium can be described by more different microstates than the system removed from equilibrium. Correspondingly, a system that is removed from thermal equilibrium has fewer microstates than when it is in thermal equilibrium and temperature is uniform throughout. When W is large, as when the system is in thermal equilibrium, the logarithm of W is correspondingly large.[14] Applying this reasoning to Eq. 3.5, the entropy (S) for a system of gas molecules is maximum when molecular disorder, W, is maximum at thermal equilibrium. From this, Boltzmann demonstrated that entropy and molecular disorder are intimately related.

The mechanical state of a system of gas molecules in a container can be specified by a detailed description of the positions and momenta of all the particles in the

---

[10]The temperature of the system of gas molecules is not uniform throughout when the system is not at thermal equilibrium.

[11]A microstate is a description of one of the possible physical configurations of all the molecules at a given time, t. A microstate is described by the position and momentum of each molecule in the system.

[12]Boltzmann's constant, K_B, has units of J/°K so both sides of Eq. 3.5 have the same units as required.

[13]In Eq.3.5 the expression log_b is read as *logarithm to the base b* of the variable or number that follows it. Typical values that are used for the base b are 10, 2 and the constant e, where $e = 2.718281828....$ When e is used as the base, instead of writing log_e (W) the expression *ln* (W) is used and referred to as the *natural logarithm* of the number W. The logarithmic function specifies the exponent to which the base must be raised to equal the number or variable that follows the *log* or *ln* symbol. Therefore *ln* (W) is the exponent to which the base e must be raised to equal W. That means that if *ln* (W) = x, then $e^x = W$. Unless otherwise specified, log typically means log_{10}. An example for the base 10 would be: log_{10} (100) = 2 because $10^2 = 10 \times 10 = 100$. More generally for any base b, log_b (W) is the power or exponent to which the base b must be raised to equal W.

[14]For example, consider how the logarithm of a number increases as the number increases: log_{10} (100) = 2, but log_{10} (1000) = 3, and log_{10} (10,000) = 4 and so on. At thermal equilibrium, the measure of molecular disorder, W, has its maximum value for that particular system of gas molecules and log_{10} (W) = x, where x is also at its maximum value for the system.

system. Typically, the probability of different potential microstates thus defined for the molecules in a system will not be equal, and the measure of molecular disorder, W, must be replaced by an expression that takes account of the different probabilities of occurrence for different microstates, in which case the entropy of the system is defined as:

$$S = -K_B \Sigma p_i \log_b(p_i), \tag{3.6}$$

where p_i is the probability of a particular microstate of the system of atoms, and the Greek letter sigma, Σ, indicates that the expression $p_i \log_b(p_i)$ must be summed for all of the possible microstates of the system. The entropy expression in Eq. 3.6 is known as the Gibbs entropy, after Josiah Willard Gibbs the mathematician and theoretical physicist who formulated it.

Comparing the expressions for entropy in Eqs. 3.4 and 3.6, we see that entropy can be defined not only in terms of heat and temperature, but also in terms of molecular disorder. If we equate the expressions for entropy in Eqs. 3.4 and 3.6 we can write:

$$\Delta S = \Delta Q/T = -K_B \Sigma p_i \log_b(p_i) \tag{3.7}$$

Equation 3.7 states that the macroscopic thermodynamic variable known as entropy, which is the amount of heat added to a system divided by the temperature of the system at which it was added, can also be understood in terms of the statistical mechanical measure of molecular disorder.[15] When heat is added to a system, disorder increases and entropy increases likewise. We can isolate the two rightmost terms of Eq. 3.7 as follows:

$$\Delta Q/T = -K_B \Sigma p_i \log_b(p_i) \tag{3.8}$$

Multiplication of both sides of Eq. 3.8 by temperature, T, then gives:

$$\Delta Q = (T) \left[-K_B \Sigma p_i \log_b(p_i) \right] \tag{3.9}$$

Both sides of Eq. 3.9 are in units of Joules.[16] We can see from this simple algebraic manipulation another perspective of Boltzmann's observation that the macroscopic thermodynamic variable *heat* added to a system, and the microscopic

---

[15]Probability is a number that ranges between 0 and 1 and can therefore be expressed as a fraction. The logarithm of a fraction is a negative number. The minus sign at the beginning of the expression on the right-most side of Eq. 3.7 therefore means that there are two negative quantities, -1 and $\log_b(p_i)$ that must be multiplied. The overall expression for entropy as a measure of atomic disorder is therefore positive, since multiplication of two negative numbers yields a positive number. Both entropy expressions in Eq. 3.7 are therefore positive quantities.

[16]K_B has units of J/°K so the right side of Eq 3.9, which is (T) $[-K_B \Sigma p_i \log_b(p_i)]$, has units of oK (J/°K) = J.

statistical mechanical variable *molecular disorder* are intimately related properties of that system. The heat added, ΔQ, is equal to the temperature at which it is added, T, multiplied by $[-K_B \Sigma p_i \log_b (p_i)]$, which is a number that quantifies the molecular disorder of the system.

Information

One of the great surprises that emerged from the independent development of statistical mechanics in the late nineteenth Century and information theory more than 70 years later was the realization that entropy and information are intimately related. *To understand the role that entropy and information play in cosmology, and the emergence of life and mind, it is important that we examine this relationship.*

The expression for Gibbs entropy in Eq. 3.6 above bears a striking resemblance in form to the mathematical expression for uncertainty in information theory, which was derived independently of any consideration of statistical mechanics or thermo-dynamics by Claude Shannon and Warren Weaver after World War II (Shannon C and Weaver W 1949). This can be written as:

$$H = -\Sigma p_i \log_b(p_i) \qquad (3.10)$$

where H is the summed uncertainty, or surprise value, associated with a string of words in a message, or a sequence of events, each of which has a unique probability of occurrence designated by p_i. Uncertainty may also be thought of as the information conveyed by the message, or alternatively the information one would need to specify the state of a system. If it can be shown that Eqs. 3.6 and 3.10 are related, then entropy, which depends on a measure of atomic disorder, can be understood in terms of information. It will be helpful to first gain an intuitive understanding of what Eq. 3.10 is saying. In this equation, H is also considered to represent the information, *I*, inherent in the occurrence of a sequence of events or words and, p_i is the probability of each word or event. Equation 3.10 therefore states that the information content of a message depends on the likelihood or probability of occurrence of its individual components. If we consider a simple message that consists of only one word or event, the occurrence of an unlikely or rare event conveys a high level of information. The high uncertainty of its occurrence endows its actual occurrence with high information content or value. On the other hand, the occurrence of a highly probable event conveys less information because it is expected. Consider the issue from the perspective of surprise value. If tomorrow morning, the sun rises in the East, you will naturally regard this as unsurprising and therefore quite uninformative. If sunrise did not occur at the appointed time, however, you would attach a great deal of significance to this. You would recognize such a surprising event as very significant or informative; and you would also want to have an explanation. Likewise, if you received a message from a friend which stated that death and taxes were the only two things that are certain in life, you would not be very surprised. Your friend would be telling you something you already know. You would say that the message contained virtually no information. If, on the other hand, you received a message from that same friend, which stated that he had won a multimillion dollar lottery (an event with a very low probability), you would consider this to be surprisingly informative. You might even be encouraged go out to buy a lottery ticket of your own.

The relationship between the statistical mechanical description of the atomic disorder of a system of gas molecules in a container in Eq. 3.6, and the description of that system provided by information theory in Eq. 3.10, can be seen from the following example.[17] We cannot possibly know and describe the state of a system of gas molecules in a container. On the other hand, we can simplify the impossibly complex analysis that would require specification of momentum and position, for each molecule in a container of gas by revisiting the example introduced earlier for the game of pool. In doing this, we reduce the number of elements of the system and also lower the dimensions of the system from three-dimensional space for a gas to the two-dimensional surface of the pool table. To further simplify the analysis, we ignore the momentum of the pool balls and consider how the transition from the initial simple state to subsequent more complicated ones affects only how the positions of the pool balls change over time. Positional complexity can be assessed by measuring how the spatial distribution of pool balls changes over time after they are hit by the cue ball. If the surface is divided into a very large number of squares, with each one just large enough to accommodate just one pool ball, initially the racked pool balls are all confined to a small region of adjacent squares. The system is highly ordered because the position of each pool ball contains all the information needed to find all the others. But as the positions of individual pool balls begins to vary after they are struck by the cue ball they spread out over a larger total area on the table surface and are less likely to be found in adjacent squares. The system becomes more disordered. The position of each pool ball no longer contains all the information needed to find all the others. The system thus evolves from a highly ordered initial state, wherein the pool balls are clustered in a small group of adjacent squares and the position of any one of them conveys all information needed to find all the others, to a more disordered state in which, eventually, the position of each pool ball conveys only the information that specifies its own position.

Some final examples will help to illustrate the relationship between information, entropy and the Second Law of Thermodynamics. Consider the diffusion of molecules that convey the scent of roses in a room. One would not expect the scent of roses that had already filled a room to spontaneously concentrate in a small region immediately surrounding the roses, thus leaving the rest of the room devoid of their beautiful scent. Spontaneous processes in the universe always proceed from an

---

[17]The relationship between entropy and information is an example of consilience, in which two apparently different aspects of nature prove to be intimately related. Other examples consist of the statistical mechanical explanation of thermodynamic variables, such as heat, in terms of molecular motion; the realization that Newton's Theory of Gravitation is a special case of Einstein's Theory of General Relativity, to name just a few.

initial state that is characterized by a low likelihood to a succeeding state that is more likely. Correspondingly, the initial state of a system that undergoes a spontaneous change has a lower level of disorder than the final state to which it evolves. Your socks do not spontaneously gather in your sock drawer. Rather they disperse in apparently random and maddening fashion. You need more *information* to find your socks when they are spread around the bedroom and laundry room than when they are neatly gathered in your sock drawer.

On the atomic scale, we have seen that entropy is a function of atomic disorder, so a spontaneous change in a system always involves an increase in entropy or disorder, the information you would need to understand the state of the system, and the information value of any single element of the system. The universe is, therefore, characterized by an inexorable increase in entropy or disorder, and a corresponding increase in the amount of information required to specify its state, as space-time evolves. This statement is characterized as the Second Law of Thermodynamics. Spontaneous processes always proceed with an increase in entropy, or the growth of disorder. A question that commonly follows this assertion is why then do life and mind emerge in the natural history of the universe, since clearly these require more ordering of matter and energy for their existence. The answer usually given is that local order may increase in the planetary breeding grounds where life and mind emerge, but the entropy and disorder of the universe increases nevertheless because there is a greater increase in entropy outside the region where order grows.

We can now examine the relationship between Shannon-Weaver entropy (H) and Gibbs entropy (S) as follows. Although the form of Eq. 3.6 looks similar to the form of Eq. 3.10, except for the presence of Boltzmann's constant (K_B) in Eq. 3.6, the Gibbs entropy and Shannon-Weaver uncertainty are not equal to each other. Rather the Shannon-Weaver uncertainty (H), which is expressed in units of *bits* in Eq. 3.10, can be viewed as the amount of information that would be needed to specify the state of a system of particles that has Gibbs entropy (S) expressed in units J/°K in Eq. 3.6. *This means that, as a system moves toward equilibrium in a spontaneous process, the entropy of that system increases and the amount of information needed to describe the physical state of the system increases accordingly.*

The Dual Nature of Light and Matter

Newton's experiments on the nature of light revealed that white light is composed of a spectrum of monochromatic colors, or light frequencies, more commonly known as the rainbow. Newton is also responsible for advancing the particle theory of light. The idea that light consists of a stream of miniscule particles was challenged by Robert Hooke and Christiaan Huygens, who proposed the alternative theory that light is a wave that varies transversely to the direction of propagation. Subsequently, Thomas Young demonstrated that light exhibited the wave property of interference. The debate concerning whether light consists of particles or is a wave provided an early portent of the amazing findings of quantum physics, which showed that not only light, but also mater, simultaneously possesses the potential to manifest as either a particle or wave.

In addition to thermodynamics, another area of rapid advancement in physics, during the Nineteenth Century, was in the field of electromagnetic theory. Fundamental discoveries were made concerning the forces of electricity and magnetism, and these findings were synthesized by the mathematical physicist James Clerk Maxwell into the electromagnetic theory of the propagation of light. By unifying two of nature's forces, electricity and magnetism, into a more fundamental force, electromagnetism, Maxwell was able to explain that light is an electromagnetic wave that propagates through space. Maxwell's equations for light also implied that the speed of light propagation is constant irrespective of the motion of an observer relative to the motion of the light. This fact gave Einstein a vital clue in the early Twentieth Century that led him to his Special Theory of Relativity, which showed that space, and time are not constant but vary with the speed of an observer. Maxwell's electromagnetic theory of light was also consistent with the earlier ideas of Thomas Young.

Newton's idea that light consists of a stream of miniscule particles was challenged by the experiments of Thomas Young, as mentioned above. In what has come to be called the double-slit experiment, Young was able to show that if a beam of light was passed through two adjacent slits in an opaque object, the two resulting beams of light that emerged from the slits would exhibit the wave property of constructive and destructive interference. To understand this, consider how two ocean waves that are converging on the shoreline from two different angles interact with each other when they meet. When the waves converge, the size of the resulting wave depends on which part of each wave meets the other. Where the peak of one wave meets the other's trough, the waves cancel each other. On the other hand, where the peak of one wave intercepts the peak of the other wave, they add to produce a peak that is more intense than either wave initially. Finally, when the trough of one wave meets the trough of the other, the waves add to produce a trough that is lower than either of the original troughs. The example of colliding wave fronts illustrates the destructive and constructive interference that Young observed for intersecting light beams when they reached a screen where they produced an image that is called a diffraction pattern. The image shows regions of more intense light and regions of minimal light as seen in Fig. 3.2.

Young's definitive experiment proved that light is a wave, but this settled things only temporarily. The debate in the time of Newton and Young, concerning whether light consists of particles or is a wave, provided an early hint of the amazing findings that would be observed when physicists began to probe matter at the atomic level. The wave theory of light was about to be challenged!

In the photoelectric effect first discovered by Heinrich Hertz in 1887, light that falls on a metal surface causes negatively charged electrons to be ejected from the metal. This is illustrated in the Fig. 3.3 below.

Capture of these electrons into a closed circuit allowed the number of ejected electrons to be measured by the strength of the electrical current in the circuit. The strength of the photocurrent, I, is determined by the number of electrons flowing

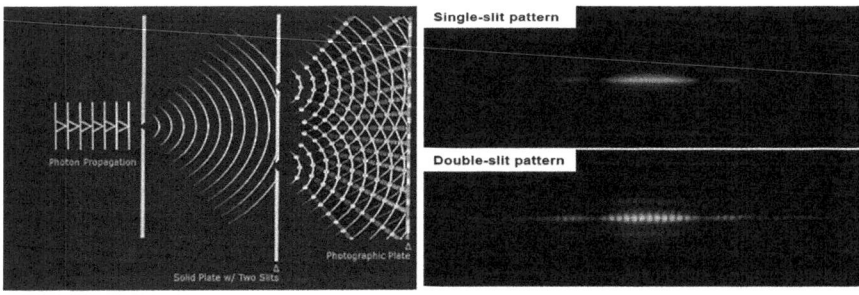

Fig. 3.2 Right. Images of red light that emerges from a single slit (top), or double slits (bottom), in an opaque object. The top image with one slit closed. The bottom image with both slits open. The separation between the two slits is 0.7 mm. The bottom image shows the signature pattern of interfering waves. The top image shows no interference pattern because only one wave emerges from the single slit. **Left.** View of Young's double-slit apparatus with both slits open. The light source on the left shines light through a single slit in the first opaque screen, after which it passes through two open slits in the second opaque screen. The bright spots that form on the photographic plate are the result of wave interference as shown. Image at right is provided by: Wikipedia at http://en.wikipedia.org/wiki/Double-slit_experiment. License: *CC BY-SA: Attribution-ShareAlike*. https://creativecommons.org/licenses/by-sa/4.0/ Image at left was obtained from an open source at: https://theobservereffect.wordpress.com/the-most-beautiful-experiment/

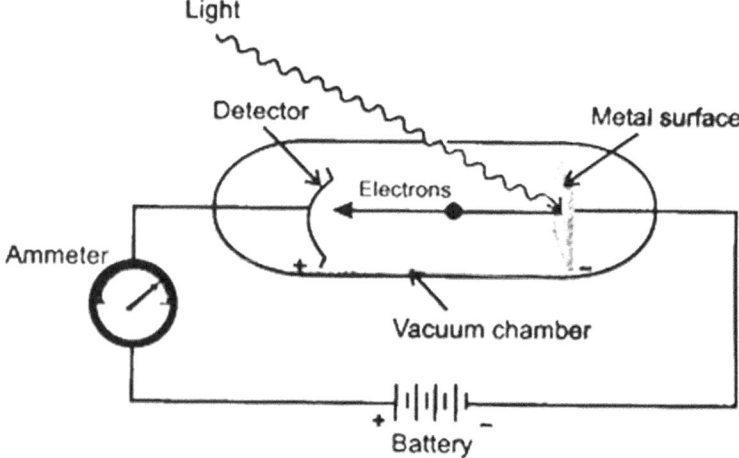

Fig. 3.3 Incident light ejecting electrons from a metal surface and the electrical circuit designed to study ejected electrons. A battery or variable voltage source applies a voltage between the metal surface and the detector. Light hits the metal surface and ejects negatively charged electrons which are captured by the positively charged detector. The Ammeter measures the current produced by the photoelectrons. The image was produced by Utkarsh Agarwal, and was obtained at: https://www.quora.com/How-can-I-understand-the-photoelectric-effect-easily. This material is reproduced here under terms of a license found at: https://www.quora.com/about/tos

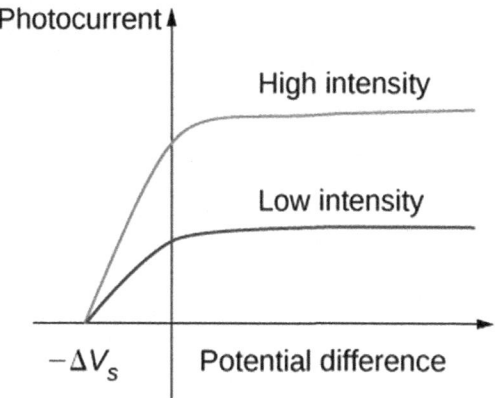

Fig. 3.4 The strength of the photocurrent produced by high and low intensity incident light, varies with the strength of the voltage (potential difference) applied between the metal surface and the detector. Photocurrent declines when the voltage approaches zero at graph origin and then becomes negative. The detector then gradually becomes negatively charged, and repels negatively charged electrons making it harder for the electrons to enter the detector. When the voltage reaches a limiting value of $-\Delta V_S$, electrons are no longer ejected from metal surface by the light and the photocurrent stops. Image was downloaded from: https://phys.libretexts.org/TextMaps/General_Physics_TextMaps/Map%3A_Universty_Physics_(OpenStax)/Map%3A_University_Physics_III_(OpenStax)/6%3A_Photons_and_Matter_Waves/6.2%3A_Photoelectric_Effect. License is at: https://creativecommons.org/licenses/by-nc-sa/3.0/us/

through the circuit, which depends on the intensity of the light that falls on the metal surface and the voltage applied between the metal surface and the detector as shown in Fig. 3.4, which shows the photocurrent produced by two different intensities of incident light.

This is in line with expectations based on the wave theory of light. According to the wave theory, however, the kinetic energy of the ejected electrons should also depend on the intensity of the light falling on the metal surface. Surprisingly, it was found that the kinetic energy of the ejected electrons depended not on the intensity of the incident light but its color or frequency. A higher frequency of light imparted more kinetic energy to ejected electrons than a lower frequency, as seen in Fig. 3.5b. This finding was explained by Einstein in a 1905 paper for which he later won the Nobel Prize. Einstein knew that Max Planck had proposed that light energy is absorbed and emitted by matter in discrete units according to the equation: $E = hf$. In this equation E is the electromagnetic energy that is absorbed, h is Planck's constant and f is the frequency of the electromagnetic wave. Einstein's insight was to apply Planck's concept to the photoelectric effect. He proposed that Planck's equation be used to determine the energy of each unit of electromagnetic radiation absorbed by the metal plate in an experimental arrangement such as that shown in Fig. 3.3. In this formulation of the collision between one photon and one electron, electromagnetic energy can be transferred to the electron only in units equal to whole number multiples of hf. It is for this reason that the kinetic energy (KE) of ejected photoelectrons depends on the frequency of the incident light and not its intensity or

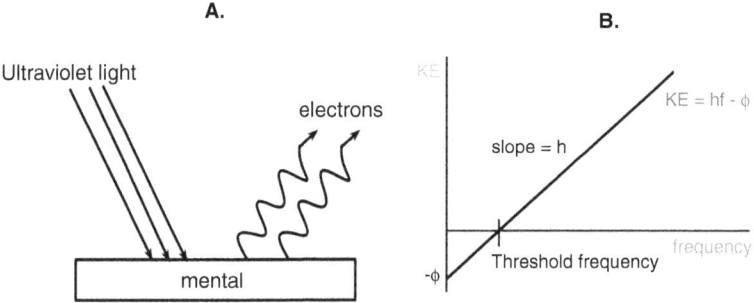

Fig. 3.5 (**a**) Illustration of the photoelectric effect. Light interacts with electrons in the metal in discrete units called photons. One photon transfers energy to one electron, which escapes the metal surface if it has a kinetic energy greater than ϕ. (**b**) Kinetic energy (KE) of photoelectrons ejected from metal plate as a function of the frequency (f) of light that hits the plate. Measurements showed that KE varied with the frequency of light not its intensity. The relationship between the kinetic energy of ejected electrons and the frequency of incident light can be expressed as KE = hf $-\phi$, where ϕ is the intercept of the KE-Axis that defines the amount of kinetic energy the electron expends escaping the metal. The image in A was obtained at: https://www.siyavula.com/science/grade-12/12-optical-phenomena-and-properties-of-matter/12-optical-phenomena-and-properties-of-matter-02.cnxmlplus under license terms found at: https://creativecommons.org/licenses/by/3.0/;. Image in B obtained at: http://dev.physicslab.org/Document.aspx?doctype=3& filename=AtomicNuclear_PhotoelectricEffect.xml and used with permission of Catherine H. Colwell the creator and copyright owner

brightness. The intensity of incident light increases the number of ejected electrons, and hence the intensity of the photocurrent as shown in Fig. 3.4, while the frequency of incident light determines the kinetic energy of the photoelectrons as shown in Fig. 3.5.

Einstein's explanation of the photoelectric effect provided support for Newton's particle theory of light. On the other hand, this particle nature of light conflicts with the wave property of interference that Young had demonstrated in the double-slit experiment. The nature of this paradox is described as the *wave-particle duality* of light, which was the first great paradox to be revealed in the quantum theory of light and matter. The next paradox encountered was even more startling, however.

In 1924, a young French Ph.D. student named Louis de Broglie reasoned that, since nature is symmetrical in so many ways, and since there is a wave-particle duality of light, perhaps there is also a wave-particle duality for atomic particles such as the electron. Using the same formula that described the wavelength of light, de Broglie predicted the wavelength of the electron and this prediction was subsequently borne out by experiments that studied the scattering of a beam of electrons aimed into a crystalline solid. Scattering would be expected if the electron was a wave. Perhaps even more surprising were the results obtained when a beam of electrons was used instead of light in the double-slit experiment. A clear interference pattern, such as the one shown for light on the right side of Fig. 3.2, was observed on the screen. This could only be explained if the electrons were moving through space as waves and not as particles. Experiments such as these established the wave-particle duality of matter. Light could behave as either a particle or a wave, and so could the subatomic and even atomic constituents of matter! If you find this baffling, you are in good company. In his Messenger Lecture in 1964, physicist Richard Feynman said (Feynman R 1967):

> I think I can safely say that no one understands quantum mechanics. . . . . .I am going to tell you what nature behaves like. If you will simply admit that maybe she does behave like this, you will find her a delightful, entrancing thing. Do not keep saying to yourself, if you can possibly avoid it, 'but how can it be like that?' because you will get 'down the drain,' into a blind alley from which nobody has yet escaped. Nobody knows how it can be like that.

Yet the mystery was to deepen even further.

Schrödinger's Wave Mechanics and the Meaning of the Quantum Wave

What does it mean to say that an elementary particle such as an electron has the properties of a wave, behaves as a wave, or even that it *is* a wave as is sometimes stated? Erwin Schrödinger developed a mathematical formulation of quantum physics called *wave mechanics*. Building on de Broglie's wave theory of matter, Schrödinger showed that particles such as the electron can be characterized by the same type of equations that describe macroscopic waves such as those that occur in water or as sound waves in the air. Schrödinger's theory was useful in helping to explain how a beam of electrons could produce a wave interference pattern in the double-slit experiment, but it left the physical meaning of the electron wave unexplained. Further information was provided by new discoveries as described below.

It was clear that the wave interference pattern could be obtained with a beam of electrons in the double-slit experiment just as it had for light. When single electrons were fired in succession toward two open slits, however, a baffling observation was made. As expected, each electron could be observed to strike the screen as a particle, but after many electrons had been fired at the double-slits the familiar wave interference pattern emerged. That is, the same interference pattern was observed after many electrons had been fired one at a time in succession as when a continuous beam composed of many electrons at once was fired! The surprise derives from the presumption that, when fired one at a time, each electron would have to travel through only one of the two open slits. In this case, no interference pattern should occur. The interference pattern would require the interaction of electron waves that passed through both of the open slits at the same time. The fact that the sequential electron firing experiment nevertheless did produce an interference pattern led to the conclusion that when electrons are fired one at a time, the interference pattern must result from the fact that each electron travels through space as a wave that simultaneously passes through both slits and interferes with *itself*. Alternatively, the passage of a single-electron wave that passes through one of the slits would have to interfere with the wave of another electron that passes through the other slit at a different time. Observations such as this begin to shake our understanding of the fabric of reality. Production of an interference pattern when single electrons are fired in sequence toward double slits would be impossible if each electron behaved as a particle in the classical sense of being located exclusively in a well-defined region of space. Instead the interpretation of the results in the sequential single electron double-slit experiment suggests that an electron not only has the properties of a particle that are directly observed through an experimental measurement that occurs at a specific place, but also has a wave nature that transcends the confines of location, and time as well. We will return to this issue in the next chapter.

Clearly, we are dealing with a paradox in the wave-particle duality of matter and light. An attempt to resolve the paradox was offered by Niels Bohr, Werner Heisenberg and others in what came to be known as *The Copenhagen Interpretation.* According to this idea, a measurement "collapses" the wave aspect of a quantum entity by virtue of which it loses the transcendent properties of non-locality. Thereafter the quantum entity manifests as a particle at the time of the and place of a measurement *by virtue of the measurement!* It was in this context that Max Born proposed that the *physical meaning* of a quantum entity's wave is that it provides information concerning the possibilities of the particle's location in different regions of space. Specifically, Born proposed that the *square of the quantum wave amplitude* for any point in space represents the probability of finding the particle there. Any observation or measurement of its wave state would cause the particle-wave to manifest exclusively as a particle *localized* at the time of the measurement in a defined region of space. The interference of matter waves, helped to establish the wave nature of matter, but the idea that a quantum entity transitions from its wave state to a particle state on the basis of a measurement that causes the quantum entity's wave to "collapse" led to questions concerning the nature of the quantum wave described by Schrödinger. Does Schrödinger's wave equation merely describe a mathematical function, or does it describe a wave that has physical meaning? Born's interpretation of Schrödinger's wave equation raised the obvious question: if quantum matter waves are described by mathematical functions that can be used to calculate the spatial distribution of a particle's probability of being localized at different regions of space, what is the *physical reality* of such waves that allows them to behave the same way in the double-slit experiment that electromagnetic light waves do? It is important to remember in this context that this is not the first time such a question has been asked. The physical meaning of the force fields of classical physics, such as the electric and magnetic fields, is defined by mathematical expressions that describe how these fields affect various elementary particles at different locations within the field. For example, two electrons will repel each other because elementary particles that both carry a similar charge exert a repulsive force against each other. This force is inversely proportional to the square of the distance between them. The nature of electric and magnetic fields puzzled physicists when they were first discovered, however. How could a non-material entity exert a force on matter? What was the *physical nature* of electric and magnetic force fields? This is the same question we are now asking about quantum fields. In time, physicists came to accept the force exerted by these fields as evidence of their tangible physical reality. It was certainly clear that moving orthogonal magnetic and electric fields produced electromagnetic waves such as visible light, X-rays, and radio waves. Although not material, light waves clearly have a tangible physical reality. The quantum matter wave is not a force field, but we can think of it as a *quantum field*.[18] Its mathematical representation describes the spatial distribution of possible locations of a particle throughout space if a measurement is made. So, the quantum field of the electron may be thought of as determining the probabilities for the particle manifestation of an electron, in various regions of space. The phenomenon of quantum tunneling, in which a particle literally vanishes from one side of an obstacle and reappears instantaneously on the other side of the obstacle without traversing the distance between the two locations, demonstrates this *quantum field* of an elementary particle very dramatically.[19] We might argue, therefore, that the *quantum field* has a real physical meaning and is as real or tangible as a force field.[20] This line of reasoning does not explain how the quantum field of an electron can manifest the property of wave interference, however. How does the quantum field of an electron interact with two slits in an opaque object? The conundrum concerning the interpretation or meaning of the quantum wave was captured by Matthew Pusey and his colleagues as follows (Pusey MF et al. 2012):

> At the heart of much debate concerning quantum theory lies the quantum state. Does the wave function correspond directly to some kind of physical wave? If so, it is an odd kind of wave, since it is defined on an abstract configuration space, rather than the three-dimensional space we live in. Nonetheless, quantum interference, as exhibited in the famous two-slit experiment, appears most readily understood by the idea that it is a real wave that is interfering. Many physicists and chemists concerned with pragmatic applications of quantum theory successfully treat the quantum state this way.
>
> Many others have suggested that the quantum state is something less than real (References. Omitted). In particular, it is often argued that the quantum state does not correspond directly to reality, but represents an experimenter's knowledge or information about some aspect of reality. This view is motivated by, amongst other things, the collapse of the quantum state on measurement. If the quantum state is a real physical state, then collapse is a mysterious physical process...

In their important paper, Pusey and his colleagues go on to show that strict information-based models cannot reproduce the predictions of quantum theory. In their own words:

> In conclusion, we have presented a no-go theorem, which – modulo assumptions- shows that models in which the quantum state is interpreted as mere information about an objective physical state of a system cannot reproduce the predictions of quantum theory. The result is in the same spirit as Bell's Theorem (Ref. omitted), which states that no local theory can reproduce the predictions of quantum theory.

---

[18]Perhaps with poetic license, we also might think of the quantum matter wave as a kind of *being field*.

[19]The phenomenon of quantum tunneling demonstrates an illusory aspect of space!

[20]This argument does not take account adequately of the very real and profound differences between the fields of classical and quantum physics, however.

If the quantum state is a real wave with physical meaning, what happens when a measurement is made that leads to the so-called collapse of this wave? An interesting theory, that would have cosmological significance if it were true, proposes that the collapse of the quantum wave generates a gravitational field that surrounds the quantum particle that manifests as a result of the collapse. (Tilloy A 2018). This is a fascinating idea because, *if it is verified*, it would demonstrate at least one tangible physical feature that lends ontological validity to the quantum wave. Tilloy's theory could provide a link between space-time, gravity and quantum mechanics. Namely, "collapse" is what generates a gravitational field, upon a measurement of the space-time permeating quantum field described by Schrödinger's equation, as well as a quantum particle at the center of that gravitational field. As Tilloy himself cautions, this theory awaits empirical verification:

> …how much should we *believe* in the model introduced here? *As theoretical physics is currently drowned in wild speculations delusionally elevated to the status of truth, a bit of soberness and distance is required.* The present model most likely does not describe gravity, even in the Newtonian approximation. It is but a toy model, a proof of principle rather than a proposal that should be taken too seriously. Nonetheless some lessons survive its ad hoc character:
>
> 1. There is no obstacle in principle to construct consistent fundamentally semi-classical theories of gravity.
> 2. Collapse models can be empirically constrained by a natural coupling with gravity.
> 3. A primitive ontology can have a central dynamical role and need not be only passive.
>
> If semi-classical theories of the type presented here can be extended to general relativity in a convincing way and if robust criteria can be found to make them less ad hoc (ref. omitted), then further hope will be warranted.

Despite Tilloy's warning about how "theoretical physics is currently drowned in wild speculations delusionally elevated to the status of truth", what he modestly calls his "toy model" implies an equivalence between properties of a quantum field and the corresponding properties in a quantum particle plus the gravitational field around it. We must leave it to the physicists to address this and to decide the ontological question regarding the nature or meaning of the quantum wave or field. Anyone who maintains that Schrödinger's wave-mechanical description of elementary quanta is merely a computational device, however, must explain how such an abstract entity that is devoid of its own ontological validity can produce the very real wave interference patterns observed in electron double-slit experiments. This question has been discussed and debated for over 90 years. Yet there is hope that progress in recent theoretical and experimental physics may resolve this issue. We can be sure of one thing, even now. Resolution of this issue will be the harbinger of a far-reaching and revolutionary new understanding of reality. For a first rate assessment of the various interpretations of quantum theory see Adam Becker's recent book titled, "What is Real" (Becker A 2018). Becker has given us a thorough account of the development of quantum theory that provides a sense of the adventure, biographical information on the cast of characters and a penetrating insight into the science as well.

Quantum Phase Entanglement and the Non-local Nature of Reality

Although he had played a major role in helping to establish quantum theory Einstein was disturbed by two of its key features. He was troubled by the role of probability in explaining quantum phenomena.[21] The other aspect of quantum theory that Einstein found objectionable was the non-continuous nature of quantum phenomena, despite his role in demonstrating the non-continuous quantum nature of the interaction of photons and electrons in the photoelectric effect. In general, the absorption of light by matter occurs in discrete non-continuous units, because electrons orbiting atomic nuclei must do so in specific orbits characterized by discrete energy levels. When the energy of a photon is absorbed by an electron in a low energy orbit the electron jumps to a higher energy orbital. There are no intermediate energy orbitals. These so-called quantum jumps violated Einstein's belief that physical phenomena like energy absorption should occur in a continuous process rather than in discrete steps. The debates between Einstein and Niels Bohr were famous and Einstein, who was well-known for conceiving thought experiments to illustrate paradoxes and conflicts in the data, was highly motivated to describe a paradox to support his belief that quantum theory was fundamentally flawed, or at least incomplete. To this end in 1935, Einstein and his colleagues Boris Podolsky and Nathan Rosen published a paper in which they portrayed what they believed was an internal contradiction that disproved quantum theory, or at a minimum showed that it represents an incomplete description of reality (Einstein A Podolsk B and Rosen N 1935).

The paradox that the Einstein-Podolsky-Rosen (EPR) thought experiment proposes would arise from certain measurements on electrons moving apart from each other. It has also been explained more recently in terms of paired photons and other quantum particles. One compelling and clear explanation derives from consideration of an electron and its anti-matter counterpart, the positron. The electron and positron are emitted from a single event at the same location after which they move apart at high speed. Quantum theory requires that they have opposite *spin*[22] owing to the fact they are a matter/anti-matter pair. Since they were created together, they have a conjoint wave function that is described as a superposition of states in which both the electron and the positron are each in the spin 1 and spin −1 state simultaneously before a measurement of spin is made. Moreover, the electron and positron do not each have a separate wave function but they share a composite one that describes all present and future possibilities of properties that they could manifest at a time and place of measurement. The measurement of either the electron's or the positron's spin forces the superposition of states wave function for that particle to "collapse" after which only one spin direction exists for that particle. It is impossible to know in advance of the measurement which spin direction will result from the collapse of the

---

[21] See discussion of Einstein's quip on this issue in Chap. 2.

[22] Spin is the intrinsic form of angular momentum possessed by quantum particles. It is assigned in multiples of integer, or half integer, units.

wave function. There is a 50/50 chance of either spin 1 or spin -1 being measured. The EPR paradox arises from the fact that whenever the spin of one of the particles is measured, the probability of spin is absolutely determined for the paired particle. It must be the opposite of the spin of the measured particle. The second particle no longer has an indeterminate spin with an equal chance of manifesting either upon measurement. When measured, the second particle will always have a direction of spin that is opposite to that measured for the first particle. The paradox arises from the necessity of instantaneous transmission of the information of the first particle's spin state to the second particle even over great distances. The instantaneous transmission of information would violate Einstein's theory of special relativity, which requires that nothing, not even information, can travel through space faster than the speed of light.

This paradox was explained by EPR more generally as follows:

> There exists a connection between the particles such that the fact of an observation of particle A is relayed to the distant particle B, in such a manner that the communication, does not diminish with distance, cannot be shielded,[23] and travels faster than light.

The fact of the two particles once being together is sufficient to mingle the particles' phases. The mingling of phases, which is known as quantum phase entanglement, derives from the conjoint nature of the wave function that describes the possible quantum states for both particles that could manifest upon measurement. The requirement of quantum physics for the instantaneous communication of information about the quantum state of one particle to its conjoint particle, irrespective of the distance between them, describes a non-local causality, or simply "non-local" aspect of reality. Ordinary light-speed-limited phenomena, on the other hand, are referred to as "local". Einstein had hoped by means of this imaginary experiment to disprove quantum theory, but subsequent experiments demonstrated that quantum phase entanglement is a true description of reality, and the theoretical work of Irish physicist John Stewart Bell showed that "all conceivable models of reality must incorporate this instant connection." Bell's Theorem is a mathematical proof that reality is non-local. This result is fundamentally important. It shows that local reality cannot be isolated from the rest of universe, which is equivalent to saying that there is a unitary nature of reality. In regard to quantum entanglement and the non-local nature of reality, David Bohm and Basil Hiley wrote:

> We bring out the fact that the essential new quality implied by the quantum theory is non-locality i.e. that a system cannot be analyzed into parts whose basic properties do not depend on the state of the whole system. We do this in terms of the causal interpretation of the quantum theory, proposed by one of us (D.B.) in 1952, involving the introduction of the 'quantum potential', to explain the quantum properties of matter.
>
> We show that this approach implies a new universal type of description, in which the standard or canonical form is always supersystem-system-subsystem. In quantum theory, the relationships of the subsystems depend crucially on the system and supersystem in which they take part. This leads to the radical new notion of unbroken wholeness of the entire universe." (Bohm D and Hiley B 1975)

---

[23]*Shielded* in the sense of being blocked or prevented.

This notion of Bohm and Hiley regarding "the unbroken wholeness of the entire Universe" is consistent with recent theoretical work which showed that the universe can be modeled accurately as the expansion of a single quantum wave function (Ernest A D 2012). The implication of this work is that all quanta in the universe are particles that are entangled in the matrix of reality that Bohm and Hiley call the "unbroken wholeness of the entire universe".

Every Picture Tells a Story

More recently, the phenomenon of quantum entanglement was demonstrated in a striking manner. As explained in Nature News (27 August 2014):

> Physicists have devised a way to take pictures using light that has not interacted with the object being photographed. This form of imaging uses pairs of photons, twins that are 'entangled' in such a way that the quantum state of one is inextricably linked to the other. While one photon has the potential to travel through the subject of a photo and then be lost, the other goes to a detector but nonetheless 'knows' about its twin's life and can be used to build up an image.

The following more detailed description is based on an article in New Scientist (Sarchet P 2014). The images shown in Fig. 3.6 are of a cat stencil, and were made using entangled photons. The photons used to generate the image at the camera never interacted with the stencil. Rather, the information used to create the image was obtained from photons that illuminated the stencil but were never seen by the camera. When photons are entangled they share a single quantum state, as explained above. Measuring the state of one of the entangled photons causes a correlated change in state of the other. The dramatic imaging study discussed here used quantum phase entanglement of photons with different wavelengths to make the images shown below without directly photographing it. Yellow and red pairs of

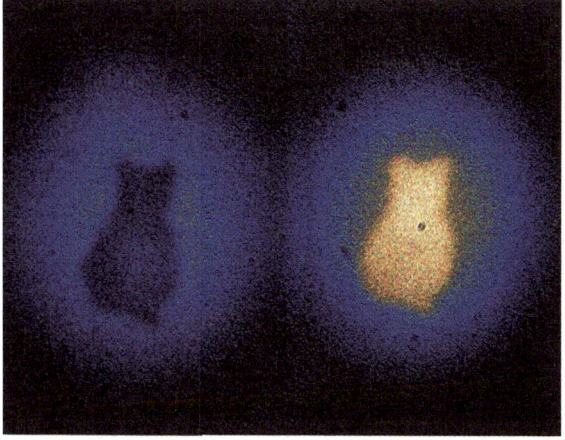

Fig. 3.6 Images of a cardboard cut-out of a cat produced by photons that never interacted with the cut-out itself, but were entangled with photons that did. Figure from (Gibney E 2014) based on the work of (Lemos GB et al. 2014). Permission to use under license number 4198331050720 obtained from Nature Publishing Group and Copyright Clearance Center

entangled photons were generated and then the red photons were fired at the cat stencil, while the yellow photons were sent to the camera. Thanks to their entanglement with the red photons, that had interacted with the image, the yellow photons had "access" to the image "information" and could form the image of the cat stencil without ever having interacted with it. The silicon stencil was transparent to red light but not yellow light and the camera was sensitive to yellow light but not red light. It is therefore impossible for the image to have been formed by photons of red light that had interacted with the image. This remarkable experiment demonstrates that quantum phase entanglement is real and can be used to image objects that are "invisible" to the photons that create the image.

In general, the contradiction that paradox reveals arises from incompleteness of understanding. A corollary of this idea is that a more fundamental explanation of the phenomena under investigation would eliminate the paradox. The EPR thought experiment reveals the paradox between the requirement of quantum mechanics for instantaneous transmission of information between two particles separated by a vast distance, and special relativity's prohibition against anything traveling faster through space than light speed. The non-local nature of reality implied by quantum phase entanglement, which appears to contradict the reality of distance between the particles, has been demonstrated many times experimentally since the EPR paper was published. Perhaps a more fundamental explanation that would resolve the EPR paradox is provided by the idea that the entanglement of all quantum particles in the universe is mediated by their "connection" to another "point" that exists outside of space-time. Such a transcendental point would therefore be in simultaneous instantaneous relation to all the quantum particles that exist in space-time. In this scenario, the requirement for information to travel through space at supra-luminal velocities is eliminated, but it becomes necessary to invoke a connectedness of everything in the universe to a "point" or reference frame that exists "outside" of it. In fact, such a point has been hypothesized by some physicists and philosophers in an attempt to resolve the EPR paradox. John Bell, for example, hypothesized that "something" might be coming from "outside space and time" to instantaneously correlate measurements of widely separated entangled particles. Toward the same end, Huw Price proposed the existence of what he called an Archimedean perspective, or point of view, that he defined as being neutral to the time asymmetry that we observe from our time-bound point of view (Price H 1996) Simultaneity of action, for two entangled quantum particles at a distance, would then be "mediated" through the Archimedean "domain". Such a domain is necessarily a higher-dimensional and transcendental one that is neither bound by time nor space; and Price even refers to the Archimedean perspective as that of God. On page 4 of his book, Price says:

> I want to show that if we want to understand the asymmetry of time then we need to be able to understand, and quarantine, the various ways in which our patterns of thought reflect the peculiarities of our own temporal perspective. We need to acquaint ourselves with what might aptly be called the view from nowhen.

To provide a perspective on "the view from nowhen" Price states on page 145:

> Consider, for example, the perspective available to God, as She ponders possible histories for the universe.

Price speculates that a version of quantum physics that incorporated the transcendental Archimedean perspective, which is necessarily symmetric with respect to time, could provide a more complete description of reality than the present version of quantum mechanics and could therefore resolve the paradox that EPR identified. A mathematical formalism that transforms the equations for the four-dimensional space-time representation of quantum mechanics to equations that account for the Archimedean, or supra- dimensional, perspective is needed to bring this idea to fruition. The supra-dimensional perspective of our four-dimensional space-time is discussed in more detail in the next chapter. We will return to Huw Price's ideas then to evaluate in more detail their importance for developing a better understanding of the phenomena of quantum physics.

Some of the Main Conclusions of Quantum Physics

We have in quantum physics the most fundamental and accurate, yet amazing and mystifying, description of reality that science has produced. Clearly when scientists contacted nature at the atomic and sub-atomic scale, some very strange and perplexing phenomena were encountered. The interpretation of these findings is still debated fiercely among philosophers and physicists alike in an effort to understand the nature of reality at its deepest level.

It is useful to summarize some of the main observations of quantum physics, which offer a description of the universe so fraught with paradox, and new insight:

1. All quantum particles, whether massless photons or particles with mass such as the electron and other subatomic particles, exist before measurement as a waveform that extends throughout space.
2. This waveform describes all the possible states that the particle could manifest if measured at a particular time and place
3. The square of the waveform's amplitude at each location describes the probability of finding the particle there upon measurement.
4. When measured, the waveform "collapses" and a particle manifests at the time and place of the measurement.
5. Both light and matter possess this wave-particle duality.
6. All the matter and energy of the universe may be part of an "unbroken wholeness", or unity of being, in which all of the constituents are in mutual instantaneous connection, irrespective of the distance between them.
7. In short there is a unitary nature of reality that supersedes, or is at least as real as, the apparent separateness of the components of reality

8. The instantaneous correlation of states that occurs between entangled quantum particles was portrayed as paradoxical by EPR because not even information can travel at supra-luminal speeds. Yet quantum entanglement has been proven to be an accurate description of reality.
9. Some notable philosophers and physicists have attempted to eliminate the EPR paradox by suggesting that all quantum particles are connected to a "point" that has its existence outside of space-time, and which provides instantaneous connectedness among all quantum particles in the universe.

It is worth reminding ourselves that these are the findings of rational empirical science. Despite the presumption of validity conferred by repeated experimental replication of results, quantum physics defines a reality that is far removed from ordinary experience, a view of reality so paradoxical that it approaches the mystical. Yet the possibility of an even deeper level of reality has been suggested. In String Theory, the idea of vibrating strings with a length near the Planck length, or 1.616252×10^{-35} m, has been advanced as a form of ultimate sub-atomic particle.[24] That is, all the other sub-atomic particles consist of these strings whose vibrational frequencies determine the type of particle that exists. According to string theory, the different properties that two different quantum particles exhibit is determined by the vibrational frequency of strings in much the same manner that the difference between two pure musical tones is determined by the frequency of vibration of the air. If the tenets of string theory are correct, and they have yet to be experimentally verified, then we must imagine that as a result of the Big Bang space-time was created and filled with quantum strings vibrating at various frequencies to produce the elementary quanta of the nascent universe.

References

Bacon F (1620) Novum Organum. p. 74. Novum ed. Joseph Devey M.A. P.F. Collier. 1902
Becker A (2018) What is real. Basic Books, New York
Bohm D, Hiley B (1975) On the intuitive understanding of non-locality as implied by quantum theory. Found Phys 5(1):93–109
Boltzmann L (1877) On the relation between the second law of the mechanical theory of heat and the probability calculus with respect to the theorems on thermal equilibrium. Sitzungsber Kais Akad Wiss Wien Math Naturwiss Classe 76:373–435
Count Rumford B (1798) An inquiry concerning the source of the heat which is excited by friction. Philos Trans R Soc Lond 88:102
Eddington AS (1928) The nature of the physical world. Cambridge Scholars Publising, Cambridge, p 69
Einstein A (2000) The expanded quotable Einstein. Princeton University Press, Princeton, p 202
Einstein A, Podolsky B, Rosen N (1935) Can quantum-mechanical description of reality be considered complete? Phys Rev 47:777–780

---

[24] 1.616252×10^{-35} m = 0.0000000000000000000000000000000001616252 m.

Ernest AD (2012) Does quantum wave packet expansion influence cosmic evolution. The Australian Institute of Physics Congress

Feynman R (1967) The character of physical law. The M.I.T. Press, Cambridge, p 129

Gibney E (2014) Entangled photons make a picture from a paradox. Nature News

Lemos GB, Borish V, Cole GD, Ramelow S, Lapkiewicz R, Zeilinger A (2014) Quantum imaging with undetected photons. Nature 512:409–412

Maxwell JC (1867) On the dynamical theory of gases. Philos Trans R Soc Lond 157:49–88

Maxwell J C (1871) Theory of heat. Dover Publications, Inc. Mineola, New York, pp 301–333, 2001

Price H (1996) Time's arrow and Archimedes' point. Oxford University Press, p. 4. Ibid. p 145

Pusey MF, Barrett J, Rudolph T. (2012) On the reality of the quantum state. arXiv:3328v3[quant ph]18 Nov

Sarchet P (2014) Schrodenger's cat caught on quantum film. New Scientist

Shannon CE, Weaver W (1949) The mathematical theory of communication. University of Illinois Press, Champaign

Tilloy A (2018) Ghirardi - Rimini - Weber model with massive flashes. arXiv.1709.03809v2 [quant - ph] 26 Jan

Chapter 4
Cosmogenesis

Richard J. Di Rocco

With Wisdom God Created the World

> Gen. 1:1 (This is the Aramaic translation of the original Hebrew text, as pointed out to me by Gerald Schroeder, author of "The Science of God".)

It is Yahweh who made the earth by His power, who established the world by His wisdom. By His understanding He stretched out the heavens.

Jer. 10:12

Nothing comes from nothing. (This is a paraphrase of a section of a metaphysical poem by Parmenides (mentioned previously in Chap. 1). The work has survived in fragmentary form, but the metaphysical section that is relevant to ontology is complete. It is discussed at length in Chap. 8.)

Abstract This chapter offers a review of the basic elements of cosmology, or what we know about the origin and evolution of the universe and how we know it. An early Static Theory about the universe, first propounded by Sir James Jeans and latter championed by Fred Hoyle, which described the universe as eternal and unchanging on the macro level, is contrasted to the Big Bang Theory, which described space and time as having a beginning approximately 13.7 billion years ago. The emergence of the Big Bang Theory as the consensus view of physicists and cosmologists is reviewed, and some more recent questions and theories in the field of cosmology are raised. Among these is the concept of an eternally cycling universe, in which each phase or cycle begins with a Big Bang and ends in a Big Crunch that is driven by the gravitational slowing of the

R. J. Di Rocco (✉)
Psychology Department, University of Pennsylvania, Philadelphia, PA, USA

Psychology Department, St. Joseph's University, Philadelphia, PA, USA
e-mail: richdi@upenn.edu

© Springer Nature Switzerland AG 2018
R. J. Di Rocco, *Consilience, Truth and the Mind of God*,
https://doi.org/10.1007/978-3-030-01869-6_4

expansion of space and its eventual collapse into a singularity that gives rise to another cycle in a Big Bang and collapse *ad infinitum*. Alternatively, the concept that our vast universe may be only one of infinitely many that collectively comprise the postulated eternal multiverse is presented. If the eternal multiverse is proven to exist, its eternal nature poses certain ontological questions such as whether it can be the sufficient cause of itself or whether a higher-dimensional transcendental cause is the source of all being.

Keywords Big bang theory · Cosmic microwave background · Cosmic inflation and the eternal multiverse · Fractal nature of the multiverse

During the early part of the Twentieth Century, the known universe consisted primarily of the Milky Way Galaxy and what appeared to be spiral gaseous nebulae that were telescopically visible in the night sky. The nature of the nebulae was intensely debated, but remained unclear. The situation was soon to change. In 1929 humanity's view of the universe was altered dramatically with the publication of Edwin Hubble's astronomical discovery concerning the real nature of the nebulae. In 1919, Hubble began his telescopic observations of the Giant Nebula that is visible in the Andromeda constellation. The images he obtained clearly showed stars amidst the gaseous nebulosity, and Hubble realized that he was seeing another galaxy far beyond the bounds of the Milky Way. Hubble found many more galaxies at increasingly greater distances from Earth. He used the relationship between a Cepheid variable star's brightness and pulsation period for determining the distance between the Milky Way and other galaxies he observed. This relationship was discovered by Henrietta Leavitt of Harvard University, who unfortunately died before she could be recognized for this seminal discovery (Leavitt 1908). Hubble's most important finding, however, was that the color of the light collected from those galaxies was shifted from the shorter wavelength or blue part of the electromagnetic spectrum toward the longer wavelength or red part. He also showed that the magnitude of the red-shift increased with increasing distance of the galaxies from Earth (Hubble E 1929). The red-shift is a Doppler effect for light just as the sound of a train horn moving away from an individual diminishes toward lower frequencies as the train moves away. The fact that the degree of red-shift increased with increasing distance from the Earth meant that the galaxies farthest away were moving away from Earth with the greatest velocity. The implication of Hubble's findings was enormous.

A Russian mathematician named Alexander Friedman had previously published paper in 1922 that predicted an expanding universe based on Einstein's theory of General Relativity. In 1927 Georges Lemaitre, a Belgian physicist and mathematician, first reported that Hubble's observations supported his own ideas about an expanding universe based on Einstein's equations for General Relativity. In that paper, Lemaitre was the first to derive Hubble's Law, which described the relationship between the distance of galaxies from the Earth and the recessional velocity of those galaxies. This relationship is illustrated below (Fig. 4.1).

A later publication by Lemaitre expanded on his earlier work (Lemaitre G 1931). Hubble's finding that the galaxies that were farthest away were red-shifted the most

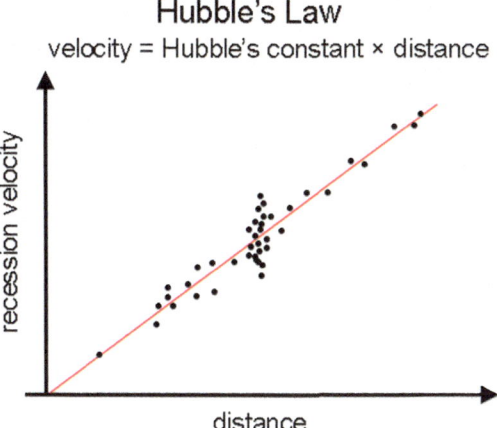

Fig. 4.1 Graph showing the relationship between recessional velocity of galaxies and their distance from Earth. Recessional velocity is based on the degree of red-shift that Hubble had observed. The linear equation that describes the relationship that is displayed graphically is also shown: $V = H_0D$ in which the slope of the line is defined by the parameter known as Hubble's Constant, H_0. This open source image was obtained at https://imagine.gsfc.nasa.gov/Images/educators/lessons/how_ far/hubble_law_plot.png and is reproduced here under guidelines found at https://https://www.nasa. gov/multimedia/guidelines/index.html

was interpreted by Lemaitre as evidence not of the movement of the galaxies away from each other through space, but of the expansion of space itself. The universe is expanding, which implies that at one time in the distant past everything was compressed into a very small region of space. On this observational basis, as well as the proposal by George Gamow that the early universe was dominated by radiation and not matter, the Hot Big Bang Theory was born.

Acceptance of the Big Bang theory of the origin of the universe was gradual, however. In the 1920s, physicist, Sir James Jeans postulated that the universe existed in a steady state. The theory was developed further by Fred Hoyle in 1948 and the question of steady state versus expanding universe persisted until the discovery of the cosmic background radiation (Penzias AA and Wilson R 1965). This discovery was decisive in convincing most theorists that the Steady State Theory was incorrect and that the Big Bang Theory represents an accurate cosmological description of a dynamic universe that had a definite beginning.

One millionth of a second after the Big Bang, the universe consisted of a hot, interacting mix of photons, electrons and baryons. Baryons are subatomic particles that are comprised of three smaller quantum particles known as quarks, which in turn are postulated to consist of the vibrating quantum strings of string theory. The mix of hot subatomic quantum particles in the earliest epoch of the universe is a fourth state of matter, after gas, liquid and solid, referred to as plasma. In this state, atoms cannot form from subatomic particles. Photons were not able to travel far in the plasma of the early universe, which would have appeared opaque as a consequence. Cooling produced by the continuing expansion of space caused the energy density of the plasma to decrease until electrons were able to combine with protons to form neutral

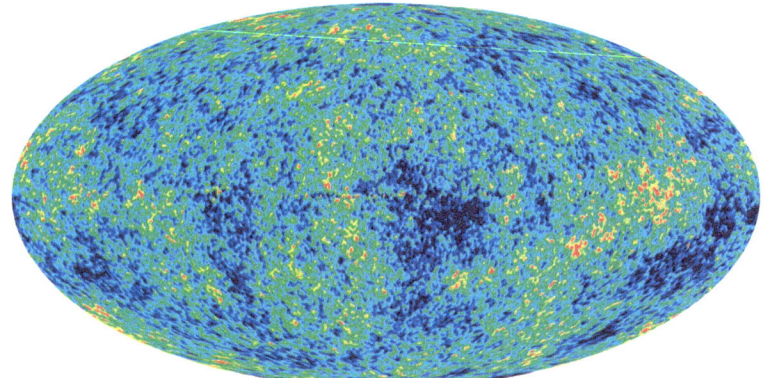

Fig. 4.2 NASA nine-year WMAP image showing the regional variation in the intensity of the CMB radiation. This open-source image was obtained at: https://map.gsfc.nasa.gov/news/. The image is reproduced under guidelines found at: https://www.nasa.gov/multimedia/guidelines/index. html

hydrogen atoms. The plasma may be thought of as partitioning or decoupling in this epoch into matter and light. This process occurred approximately 370,000 years after the Big Bang. At this point, photons were able to travel freely through space owing to a much lower rate of absorption by electrically neutral atoms compared to plasma. Those photons are detected now as the cosmic microwave background (CMB), which has a fairly uniform intensity when measured in any direction from Earth. Nevertheless, slight regional variations in the intensity of the CMB referred to as anisotropies have been measured. The latest measurement of the CMB by the NASA's Wilkinson Microwave Anisotropy Probe (WMAP) has accomplished this with a high degree of resolution as shown in the Fig. 4.2.

As mentioned above, the subtle fluctuations in the CMB were imprinted on the universe when it was about 370,000 years old. The imprint reflects quantum fluctuations of the density of the plasma in small regions of space that arose as early as the first nonillionth of a second (10^{-32} s) and then became magnified by the expansion of the universe. Gravitational attraction in regions that had a greater density of matter then gave rise to the present vast web of galaxies, galaxy clusters and super clusters that we presently observe.

The Second Law of Thermodynamics Applied to the Universe as a Whole

The Second Law of Thermodynamics applies to the universe on a cosmological scale, as well as chemical reactions in a test tube and quantum events on the Planck scale. There is an indisputable and intimate connection between the expansion of space and the inexorable growth of entropy required by the Second Law of Thermodynamics. That is, the universe's total entropy must be increasing as space-time evolves. Equivalently, we may say that as space-time evolves the total information that would be required to specify its state also increases. What accounts for this growth of entropy as space-time evolves? There are more possible configuration states of energy and matter at later stages of the evolution of space-time than earlier ones, because space expansion creates more potential locations to accommodate the distribution of a constant amount of matter and energy. If the universe is destined to expand forever, and it is a closed system that does not exchange matter, energy, or information with anything outside of itself, it would become a vastly extended entropic domain with an ever-increasing amount of information required to specify its state. If on the other hand the universe is not a closed system, as is suggested by the Big Bang itself, then the possibility of at least one exit point through which matter, energy and information-entropy leaves the universe must be considered. Could such an exit mechanism be provided by the many massive black holes at galaxy centers throughout the universe? The fate of information that falls into a black hole is an important question. According to the view of theoretical physicist Leonard Susskind it is neither destroyed nor does it vanish. Rather, it continues to exist on the surface around the black hole defined by the event horizon (Susskind L 2008). What happens to the matter and energy that fall into a black hole? Could the singularity of a black hole spawn a daughter universe outside of our own? Alternatively, could the expansion of our own universe eventually reverse into a phase of contraction that would lead to another super dense hot state laden with an immense quantity of information, referred to as a Big Crunch by some? Would it then be possible for the singularity of the Big Crunch to bounce back as a new Big Bang, and so on, in a never-ending cycle, as postulated in "Endless Universe" (Steinhardt PJ and Turok N 2007)? In "Cycles of Time", mathematician and theoretical physicist Roger Penrose discounts the likelihood of black holes spawning new universes, or our own universe being part of a cyclic one that involves a never-ending recurrence of big bangs and big crunches (Penrose R 2011). Instead he makes a case for a Conformal Cyclic Cosmology (CCC) in which our universe is one of a potentially eternal and infinite[1] sequence of universes, each of which comprises what Penrose calls an Aeon that is spawned from the "previous" one, *ad infinitum. CCC* is an interesting alternative to the cosmologies mentioned above. Penrose argues that the far distant future of one exponentially expanding universe represents a conformal boundary that is simultaneously the initial Big Bang state of the next universe or Aeon.

The Theory of Cosmic Inflation and the Eternal Multiverse

The theory of cosmic inflation, which postulates a rapid and energetic expansion of space after the Big Bang, is described by Brian Greene as follows (Greene B 2011):

---

[1]Eternal means without beginning or end, while infinite means unlimited size or number. It would be possible for an infinite chain of universes, or multiverse, to exist without being eternal if the chain had a beginning. An eternal infinite multiverse would consist of an infinite number of universes, which collectively have neither beginning nor end.

The mathematics of the Big Bang shows that in the very early universe, gravity could act in reverse. This "repulsive gravity" would repel everything around it, causing a huge expansion. This force was so powerful it could take space as tiny as a molecule and blow it up to the size of a galaxy in billionths of a second. And all that energy was instantly transformed into matter. This expansion is called "Inflation" and it was the "bang" in the Big Bang.

The mathematics of inflation suggests that there's always some part of space that is still inflating. In this picture, the Big Bang is not a unique event—multiple bangs happened before ours and countless others will happen in the future. The idea is termed "Eternal Inflation.

As Brian Greene suggests, some cosmologists believe that an inflationary Big Bang provides evidence for the existence of an eternal multiverse that consists of an infinite number of "bubble universes" that are created by the repulsive gravity in a region of a pre-existent universe. The spatial and temporal evolution of a chaotic, self-reproducing inflationary multiverse has been explained in detail by Andre Linde (Linde A 2016). Our local universe appears to be homogeneous (see Fig. 4.2), but beyond the horizon of light visible from Earth, the structure of the multiverse is complex. Pocket universes may have physical laws that differ from those of the universes that give rise to them. They may even differ in dimensionality.

The related concepts of Cosmic Inflation and the Eternal Multiverse are not universally accepted, however, owing to a lack of definitive proof. Cosmologists are searching for supporting evidence by evaluating what can be discerned about the universe (or event) that may have spawned our universe from various analyses of the CMB. Multiple groups are looking for an imprint of gravity waves on the CMB. These gravity waves are predicted, by the theory of Cosmic Inflation and Einstein's Theory of General Relativity, to have been generated during the postulated cosmic inflationary epoch. This cosmic expansion of space is hypothesized to have begun as early as 10^{-37} s after the start of the Big Bang, and to have progressed at supra-luminal speed. In March of 2014, one group that had been working with the BICEP2[2] telescope in Antarctica announced data that was interpreted as evidence of gravity waves. The BICEP2 team claimed to have identified a specific pattern of polarization, called B-Mode, of the CMB electromagnetic radiation that would have been caused by gravity waves during the inflationary epoch of the universe. These results were widely acclaimed initially, because if confirmed they would have potential to support a theory of quantum gravity, as well as confirm the inflation theory of the early cosmic expansion. Since the original announcement, however, evidence from other groups using the Keck Array and the European Space Agency's Planck telescope showed that the electromagnetic polarization that was reported by the BICEP2 team was caused by foreground galactic dust in the Milky Way. At the press conference that announced the BICEP2 results Drs. Andrei Linde and Alan Guth, who are founders of the Theory of Cosmic Inflation, made the argument that if inflation is experimentally verified it would be difficult to exclude the possibility that

---

[2]This is the acronym for Background Imaging for Cosmic Extragalactic Polarization 2.

the infinite multiverse is real. The science news magazine, "New Scientist" reported in regard to this (Grossman L 2014):

> "If inflation is there, the multiverse is there", said Andrei Linde of Stanford University in California, who is not on the BICEP2 team and is one of the originators of inflationary theory. "Each observation that brings better credence to inflation brings us closer to establishing that the multiverse is real."

> The simplest models of inflation, which the BICEP2 results were alleged to support, require a particle called an inflaton to push space-time apart at high speed. "Inflation depends on a kind of material that turns gravity on its head and causes it to be repulsive", says Alan Guth at the Massachusetts Institute of Technology, another author of inflationary theory. Theory says the inflaton particle decays over time like a radioactive element, so for inflation to work, these hypothetical particles would need to last longer than the period of inflation itself. Afterwards, inflatons would continue to drive inflation in whatever pockets of the universe they inhabit, repeatedly blowing new universes into existence that then rapidly inflate before settling down. This "eternal inflation" produces infinite pocket universes to create a multiverse.

Studies of the CMB to date do not provide definitive evidence in support of the inflationary theory of the Big Bang, but various satellite and ground-based experiments have continued working on measurements of the CMB polarization. Forthcoming data from these other measurements have the potential to demonstrate an imprint of gravitational waves on the CMB. According to Linde and others this would establish the theory of cosmic inflation and the multiverse. Confirmation would have far reaching implications, not only for physics and cosmology, but also for theology and philosophy. If the multiverse exists it is infinite and eternal, and the Consilient Truth that can be known about it must also be infinite and eternal! To say this but no more again begs the question: in what does eternal truth have its existence, or in what does it subsist?

These are heady times for cosmology, physics and humanity as science continues to probe the cause of the universe in which we live. As mentioned several times already, this is extraordinary by the standards of empirical science because scientific conclusions ultimately must be experimentally verified, and that necessarily involves experimentation *within* the universe. Study of a transcendental, or extra-universal, cause of the universe[3] has been viewed as an exercise in metaphysics. This perspective has prevailed among most scientists because, while science can literally look back in time through the Hubble Space telescope and other instruments that collect electromagnetic radiation and information from the very early universe these instruments cannot make *direct* contact with anything that exists beyond the boundary of the observable universe. Yet we see in the power of mathematics, as exemplified in theoretical physics, the potential to make inferences about transcendental events and factors outside of our universe. Moving beyond the models and predictions enabled by mathematics, we see from the BICEP2, Keck Array, Planck and other

---

[3]If we define *universe* to mean everything contained within the horizon of light that we can observe, it is possible that a region of space-time beyond this horizon was the source of the event that caused our Big Bang.

measurements that it may be possible to obtain direct experimental observations of the CMB that would allow inferences about the postulated inflationary epoch of the very early universe. A better understanding of the physics of the complex inflationary regime may eventually provide further insight into the cause of inflation itself. Since cosmic inflation is thought by many cosmologists to provide the mechanism for the Big Bang, understanding the cause of inflation comes close to understanding what caused our universe to spring into being.

While it is highly unlikely that we will ever directly observe anything outside our universe, imprints on the CMB left by events that occurred only 10^{-37} s after the Big Bang may one day allow us to make reasonable inferences about its transcendental cause. Will observations and analyses of imprints left on the CMB be sufficiently rigorous to convince the scientific community about the nature of the transcendental cause of our universe, or our region of the multiverse if indeed the multiverse is confirmed by proof of the Inflationary Theory of the Big Bang, or by some other means? If the theory of cosmic inflation is correct, then an inflaton from another universe, or a region of our own universe beyond the "horizon" that we can observe, initiated our universe. If so, we must also consider the question whether the multiverse is infinite in both past and future temporal domains? In other words, as mentioned above, did the infinite multiverse have a beginning or is it truly *eternal* in the sense that it has neither beginning nor end? Finally, in the case that the multiverse is eternal, we must ask whether it is the sufficient cause of itself, or whether it has a cause that is distinct from it. We will consider these questions further in Chap. 8.

For the present, it is interesting to note that while an eternal multiverse as a whole would manifest the eternal aspect of Sir James Jeans' and Fred Hoyle's Steady State theory, it also would possess the dynamic aspect of the Big Bang theory. This dynamic aspect would be manifested an infinite number of times in the creation of an endless temporal proliferation of "bubble universes" that are spawned when an inflaton causes a region of space to inflate or expand faster than the speed of light. To future observers of this event who reside within a newly formed universe, the inflation event would appear just as the Big Bang appears to us.

How would the eternal generation of new universes that collectively comprise the multiverse appear from the perspective of a transcendental or higher dimensional domain beyond the bounds of space-time; that is, from the Archimedean perspective advocated by Huw Price (Ibid)? We can begin to examine this question by using a computer simulation of an increasingly magnified fractal geometric shape such as the Mandelbrot Set as a model for the spatial and temporal dynamics of the multiverse. A fixed frame of this simulation is shown in Fig. 4.3.

Using a fractal geometric model for the multiverse is reasonable for us to consider because this is precisely how Andre Linde describes the multiverse (Linde A 1994). In reference to the theory of eternal inflation, Linde states:

> From this theory, it follows that if the universe contains at least one inflationary domain of a sufficiently large size, it begins unceasingly producing new inflationary domains. Inflation in each particular point may end quickly, but many other places will continue to expand. The total volume of all these domains will grow without end. In essence, one inflationary universe sprouts other inflationary bubbles, which in turn produce other inflationary

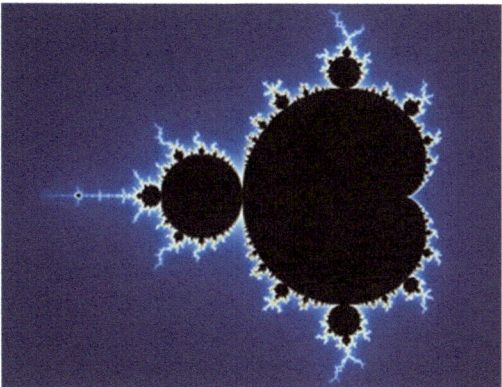

Fig. 4.3 Graphic of the Mandelbrot Set, a fractal geometric object in which all regions of the border are self-similar under magnification. The self-similar Mandelbrot fractal shapes are visible in varying sizes along the border of this still frame. Continuous magnification of the border in any of these regions reveals more self-similar fractal objects without limit. The Mandelbrot Set, therefore, contains an infinite number of self-similar fractal objects. You may view the continuous magnification of the Mandelbrot set at: https://commons.wikimedia.org/wiki/File:Mandelbrot_ sequence_new.gif. This Mandelbrot zoom sequence was produced by Simpsons contributor at English Wikipedia and released into the public domain via Wikimedia Commons

> bubbles. . . .*This process, which I have called eternal inflation, keeps going as a chain reaction, producing a fractal-like pattern of universes. In this scenario the universe as a whole is immortal.* Each particular part of the universe may stem from a singularity somewhere in the past, and it may end up in a singularity somewhere in the future. There is, however, no end for the evolution of the entire universe. [Emphasis added]

A feature of fractal geometric objects like the Mandelbrot Set, that is essential to its application as a model for an eternal self-replicating multiverse, is the fascinating property of self-similarity. This defining property owes its name to the inexhaustible supply of fractal objects along the border region under zoom (continuous) magnification. A computer simulation of a zoom magnification of any region of the border of a fractal geometric object reveals new fractal objects as the magnification progresses. In the analogy proposed here, the continuous magnification corresponds to the passage of time and the newly revealed self-similar border regions of the Mandelbrot Set correspond to the spatial dimensions of the newly formed bubble universes. The simulation shows how we can imagine the infinite[4] growth of the eternal[5] multiverse from our own perspective within space-time. On the other hand, from the perspective of a higher-dimensional or transcendental realm of a being that exists outside of space-time, the view necessarily would be different. In this case, the higher dimensional being would perceive all temporal and spatial phases of the infinite and eternal evolution of the multiverse simultaneously in what would be

---

[4]*Infinite* means without limit, such as the number of universes in the multiverse cannot be counted.

[5]*Eternal* means without beginning or end

essentially one freeze-frame in that transcendental higher-dimensional reality. We can get an imperfect or limited glimpse of what it would be like to experience this from an unlikely and surprising source. The following is quoted from a letter of Mozart (Hadamard J 1945):

> Once I have my theme, another melody comes, linking itself to the first one in accordance with the needs of the composition as a whole: the counter-point, the part of each instrument, and all these melodic fragments at last produce the entire work. . . . . . .*then my mind seizes it as a glance of my eye a beautiful picture. . . . . .it does not come to me successively. . ., but it is in its entirety that my imagination lets me hear it. [Emphasis Added].*

We will discuss Mozart's description of his creative process in a different context in Chap. 7. For now, the quote provides an opportunity to imagine what it might be like to experience sequential events simultaneously.

Comparison of Intra-and Extra-Dimensional Perspectives

We can gain further insight into the perspective of events in a realm of a given dimensionality from within that reality, as well as from higher-dimensional realities as follows. First consider how a *being* that exists in a two-dimensional line universe, which consists of one spatial and one temporal dimension, perceives sequential points along the line. This *being* necessarily experiences or perceives the points in sequence by moving from one to the other along the line in time. This is illustrated in Fig. 4.4a. Now consider how an observer living in a higher three-dimensional universe that consists of two spatial and one temporal dimension that includes the line could perceive the points labeled 1–5 Fig. 4.4b. From the observer's supra-linear perspective from within the plane, it is clear that the points along the line are perceived simultaneously. Next imagine a being that lives in a four-dimensional world that consists of three spatial and one temporal dimension. This is the four-dimensional space-time of Einstein in which we exist, and in which the multiverse is postulated to evolve. We are able to "look down" on a two-dimensional plane from our higher-dimensional perspective, just as you are doing right now by viewing Fig. 4.4b on the page. You not only see the line and the points drawn on the plane at a glance, but also other more complex geometric objects such as the rectangle. The observer in the plane, however, can only see the two closest sides of the rectangle.

Now consider how we perceive solid geometric objects in our four-dimensional space-time realm from a single vantage point. Just as the being that lives in the plane has incomplete perception of the square in Fig. 4.4b from a single vantage point, we have a limited perception of a three-dimensional geometric shape, such as a cube viewed from a single vantage point, as shown in Fig. 4.5.

Here then is the key question. How would the solid geometric shape be perceived by a being that exists in a five-dimensional world of four spatial dimensions and one temporal dimension? If we extrapolate from the examples already given, we must conclude that the higher-dimensional being can perceive all sides of the solid object

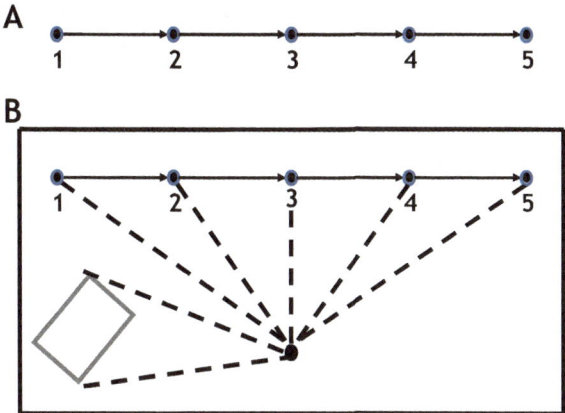

Fig. 4.4 (a). Schematic of a two-dimensional linear universe that consists of one spatial and one temporal dimension. In this universe, an observer can only perceive the positions labeled 1–6 by moving through them sequentially along the connecting line from left to right. (**b**) A three-dimensional world that is comprised of two spatial dimensions that form the plane and one temporal dimension. Notice that this three-dimensional world contains the two-dimensional one. In this universe, an observer can perceive all the positions 1–5 simultaneously, but only the two closest sides of the square. To perceive the whole square the observer confined to the plane must move to another position to change perspective in order to perceive the sides of the square at different moments of time after moving. From our perspective in four-dimensional space-time, however, we can see all the square's sides at once by simply looking down at the square on the page

Fig. 4.5 View of a cube. From our perspective within four-dimensional space-time, we only can see a maximum of three sides of a cube from any location without moving. From a higher five dimensional perspective that consists of four spatial dimensions and one temporal dimension, a being would see all six sides of the cube simultaneously

simultaneously without changing position! We have trouble "visualizing" this perspective because it is so alien to our experience, but the examples above show that all sides of a solid geometric shape must be perceived simultaneously by a higher-dimensional being. Extending the analogy, we can imagine that a transcendental higher-dimensional being also perceives all temporal phases of the evolution of the multiverse in a single "glance just as we perceive a freeze-frame of the infinite zoom magnification of the Mandelbrot set. This is so difficult for us to imagine precisely because we exist in space-time rather than beyond it in an Archimedean reality with its higher-dimensional perspective. From our perspective in space-time, we can perceive multiple points in space simultaneously, but only one moment of time at a time.

With this background, we can return to the question of the strange results of the sequential single electron double-slit experiment, which revealed the emergence over time of the same interference pattern that is produced when a beam of multiple electrons is projected toward double slits. Spreading the electrons out singly over time does not alter the results of the electron double-slit experiment. To understand this, we must appeal to Huw Price's idea of the "view from nowhen". That is, the view from the higher-dimensional Archimedean perspective. As explained above, this is a point of view that we can never experience directly but perhaps it can be understood if we acknowledge that there is a super-dimensional reality in which the sequential single electron detections at the screen in the double-slit experiment occur and are experienced simultaneously, much as Mozart says that he experiences his symphonic composition when his "mind seizes it as a glance". From the perspective of this super-dimensional reality, time is not experienced as a sequence of moments, but simultaneously as an "eternal now". Somehow, on the basis of a "projection" of this "higher" reality onto our own reality, the single electron double-slit experiment is equivalent to the many electron beam double-slit experiment. In a sense, therefore, the equivalence of the single-sequential and many-simultaneous electron double-slit experiments demonstrates an illusory aspect of time. The phenomena of quantum tunneling and quantum entanglement demonstrate the illusory aspect of space in the same way that the sequential single-electron double slit experiment demonstrates the illusory aspect of time! It seems that quantum mechanics is consistent with a perspective from a higher dimensional reality in which all of space-time is experienced "at a glance", and that what we experience is a "projection" from that higher-dimensional reality onto our own. The perspective from that higher-dimensional reality appears to involve a view not only from *no when*, but also from *every-where*!

Going beyond the issue of a transcendental or higher-dimensional perspective of our four-dimensional space-time reality, we must pursue the question concerning the potential transcendental cause of our universe. It is quite interesting that some philosophers, physicists and cosmologists have advanced the idea that the universe, or our aeon of the putative multiverse, is a virtual reality or that its ultimate destiny will be a virtual reality (Tippler FJ 1994). Now, a virtual reality necessarily implies an actual reality[6] that exists beyond its bounds, and a causative intelligent agent that exists in that true reality. Moreover, the possibility of an infinite progression of simulated realities has been raised (Bostrum N 2003). In this scenario, intelligent beings "emerge" in at least some simulated realities according to their laws and eventually develop the computational capability required to simulate another reality with this process continuing *ad infinitum*. This scenario begs the question

---

[6]*Actual reality* is a redundant expression and *virtual reality* is an oxymoron, but everyone knows what they mean. Considered together, these terms define the concept of relative realities. There is no *a priori* reason to believe that one reality is more or less "real". They are real in relation to each other as are the aeons of the multiverse of which Roger Penrose, Andre Linde and others speak. The cause of one may be found in another, and it in turn may cause yet another.

concerning how the first "reality" is initiated, however. A related concept has been advanced by Seth Lloyd, who has suggested that the universe is a cosmic quantum computer that computes its own evolution (Lloyd S 2006). You may have noted already that the so-called true reality and virtual reality that we are discussing here are reminiscent of two successive aeons of Roger Penrose's Cyclic Conformal Cosmology, or two successive universes in Andre Linde's Multiverse.

I suspect that most readers would agree that the idea that the universe is a quantum computer that computes dynamic states of itself, and the possibility that the universe is a virtual reality created by some "other being" who exists outside of it, are notions that are no less fantastic than the idea that a Supreme Intelligence expresses His potentiality or thought in the creation of the universe-multiverse. That is, these other notions are no less fantastic than the common idea that "God created the universe".[7] In keeping with this perspective, Physicist Bernard Haisch has suggested that God embodies creative potential that is actualized or manifested as the reality we observe in this universe-multiverse (Haisch B 2009). In any event, if one day it is proven that the universe is a self-programming quantum computer that creates a virtual reality for a presumably intelligent transcendental being, then the case will have been made for an intelligent designer.[8] If this view of the universe is correct, one must ask what kind of mind is capable of generating a reality, virtual or otherwise, that could spawn life from non-living matter, as well as intelligent mind that has been able to probe the origin and meaning of "reality" and "being". This is especially true given that all of this is accomplished in apparently random fashion owing to the action of entropy as the agent of natural selection (see next chapter).[9] Science is a long way from proving that the universe is a quantum computer that computes dynamic states of itself to create a virtual reality for the pleasure of a transcendental intelligent being. More to the point science may never accomplish such a feat, but we can begin to see more clearly a convergence of understanding at the interface of science, philosophy and theology. Can mathematics, philosophy and theology advance where empirical science cannot, and most scientists dare not, tread? We must now consider whether the universe itself responds to the ontological question posed by its existence by giving rise to life and intelligent mind to consider the problem, and perhaps one day to provide the definitive answer.

References

Bostrum N (2003) Are We Living in a Computer Simulation? Philos Q 53(211):243–255
Greene B (2011) www.sciencecafes.org/media/downloads/Episode_Multiverse.pdf

---

[7]Yet we find that these hypotheses are proposed by scientists, not theologians.

[8]The Engineer, or Programmer, Who exists beyond the bounds of the universe.

[9]How paradoxical that the highly-ordered states of matter required for life and mind arise as a response to the most disorganizing agent in the universe, entropy! It is a cause for wonder that physical law, forces and constants have the precise values needed for that purpose.

Grossman L (2014) Multiverse gets real with glimpse of big bang ripples. New Scientist, March 18

Hadamard J (1945) Essay on the psychology of invention in the mathematiical field. Dover Publications, Inc., Mineola, p 16

Haisch B (2009) In: Haisch B (ed) The god theory. Red Wheel/Weiser, San Francisco

Hubble E (1929) A relation between distance and radial velocity among extra-galactic nebulae. PNAS 15(3):168–173

Leavitt HS (1908) 1777 variables in the Magellanic clouds, vol 60. Annals of Harvard College Observatory, Cambridge, p 87

Lemaître G (1931) Expansion of the universe, the expanding universe. Mon Not R Astron Soc 91:490–501

Linde A (1994) The self-reproducing inflationary universe. Sci Am. 271 5 pp 48–53

Linde A (2016) A Brief History of the Multiverse. arXiv:1512.01203v2[hep-th]. pp 1–18

Lloyd S (2006) Programming the universe. Vintage Books, New York

Penrose R (2011) Cycles of time. Alfred Knopf, New York

Penzias AA, Wilson RW (1965) A measurement of excess antenna temperature at 4080 Mc/s. Astrophys J 142:419–421

Steinhardt PJ, Turuk N (2007) Endless universe. Doubleday, New York City

Susskind L (2008) Black hole wars. Little Brown and Company, Boston

Tippler FJ (1994) The physics of immortality. Doubleday, New York. pp 108–9, 208–9, 212–13

Chapter 5
Abiogenesis: The Emergence of Life from Non-living Matter

Richard J. Di Rocco and Edgar E. Coons

God formed man out of the clay of the ground.

Genesis 2: 7

We know that the whole creation has been groaning in labor pains until now

Romans 8:22

Abstract The field of abiogenesis explores the mystery concerning how life arose on Earth from non-living matter. We must assume that this happened in accordance with the laws of physics and chemistry. How chemical evolution led to the appearance of self-replicating polynucleotides has not been determined, but significant progress has been made. Uncertainty likewise remains concerning how the enzyme-driven reactions of metabolism arose, as well as how metabolic reactions and polynucleotide-based genetic mechanisms of inheritance were encapsulated together within the membranes of the first cells. This chapter examines a broad outline of molecular biology, and how this information provides retrospective insight into life's beginning by asking what kind of beginning is consistent with life's current state of affairs. Information gleaned from this exercise is then integrated with information derived from a prospective exercise that asks how self-replicating polynucleotides, could have formed from simpler organic building blocks on the basis of first principles. One of the great problems of abiogenesis is captured in the question concerning whether genes or metabolism came first. Arguments have been made in favor of independent

R. J. Di Rocco (✉)
Psychology Department, University of Pennsylvania, Philadelphia, PA, USA

Psychology Department, St. Joseph's University, Philadelphia, PA, USA
e-mail: richdi@upenn.edu

E. E. Coons
Psychology Department, New York University, New York, NY, USA
e-mail: eec1@nyu.edu

69

beginnings, but this provides no insight regarding how they were integrated in cells to yield the modern state of affairs. How would it all come together? A third possibility requires that the genes needed for the production of enzymes that catalyze metabolic reactions, and metabolism that provides the energy necessary for the production of enzymes, co-evolved from simpler antecedent processes that were always integrated.

Keywords Pre-biotic earth · Molecular biology · Self-replicating polymers in Abiogensis · Clay as catalyst in synthesis of organic polymers

Preconditions for the Emergence of Life

It is perhaps self-evident that any understanding of life and mind must include consideration of, and be consistent with, the physical laws that provide the context and causes of their emergence and evolution. What physical laws provided the impetus for life to arise initially from inanimate matter; and what physical laws provided the selective pressure necessary for the evolution of living matter once it was established? The question regarding the mechanism of the emergence of life must first be addressed in terms of permissive factors. That is, what are the necessary preconditions for life? Among these factors are the often-cited precise values of certain physical constants that are necessary for the emergence of life. For example, if the force of gravity were too high, the nascent universe would not have been able to continue expanding for long after the Big Bang, whereas if it were too low galaxies and stars would not have been able to coalesce under its influence. Without stars, the elements heavier than helium could not have been formed by stellar nuclear synthesis and carbon would not exist. Without carbon's unique properties, the complex chemistry of life would be unlikely to emerge, even from related elements like silicon. There is more, but my purpose is not to create a case for intelligent design with this information, but rather simply to note that the universe in which we exist clearly has physical laws that are necessary for the emergence of life.[1] In addition to the required values of certain physical constants, other factors are necessary. These include time, and environments conducive to mechanisms of inorganic[2] and organic[3] chemical reactions capable of generating self-replicating polymeric[4] molecules. Such conditions existed on the primordial Earth and may also exist on Earth-like planets that orbit other stars throughout the universe.

---

[1] The idea that we live in a universe that has physical laws that are compatible, or even necessary, for the emergence of life is known as the *Anthropic Principle*.

[2] Inorganic molecules do not include carbon as an atomic constituent.

[3] Organic molecules all contain carbon as one of the atomic constituents. Sugars, fats, proteins and the nucleic acids DNA and RNA are all examples of organic molecules.

[4] A polymer is a complex molecule made of many identical or similar subunits linked together. Various types of clay are examples of inorganic polymers.

Several factors are required to establish stable, high-fidelity, self-replicating polymeric molecules like RNA and DNA, as well as complex high molecular weight structural and functional polymeric molecules like proteins. These factors are: (1) mechanisms of chemical evolution that lead to the formation of organic molecular building blocks of more complex polymers; (2) energetic mechanism(s) to initiate and accelerate chemical reactions that synthesize large polymers from smaller subunits; (3) a source of energy to drive these endergonic[5] reactions forward; and (4) metabolic mechanisms to capture energy from the environment to do the biological work that is required to maintain the order inherent in more complex and less stable polymers once they are formed. Further, in keeping with the idea that causation necessarily depends on antecedent events, we can say that everything that happens in chemical and biological natural history is derived from something that preceded it. On this basis, it is reasonable to assume that spontaneous generation of life from non-living matter happened during one epoch of chemical evolution in the early history of the Earth. Considering the fact that catalytic mechanisms, metabolism and the synthesis of complex polymeric molecules, such as polynucleotides and proteins, co-exist in dynamic interdependence in cells today it is likely that they co-evolved from very early stages of chemical evolution that led to the emergence of life. It is certainly difficult to imagine how the evolution of catalysis, energy metabolism, polymeric proteins and self-replicating polynucleotides could each proceed independently to an advanced stage of development and then, at a much later stage, become synergistic and interdependent to achieve the modern state of affairs. To ignore this difficulty begs the question: how would it all come together in the independent evolution scenario? The co-evolution scenario for abiogenesis offers the alternative argument that the interdependence of catalytic mechanisms, polynucleotide synthesis, and energy metabolism that is evident in cells today arose from interdependent antecedents in the early stages of chemical evolution that resulted in abiogenesis. What would the co-evolution scenario look like? This is perhaps the most difficult, and foremost, question in Biology; and it remains unanswered, although not for lack of effort devoted to the problem.

Jim Baggott describes a scenario that involves a catalytic role for iron-nickel-sulfur (Fe_5NiS_8) containing minerals trapped in tiny pores of alkaline hydrothermal vents on the ocean floor (Baggott J 2015). In this scenario, the catalysis of carbon-fixation reactions between molecular hydrogen (H_2) and carbon dioxide (CO_2), that bubble up through the vents, occurs when these gases pass over the catalytic minerals of the vent pores. These reactions are postulated to produce a number of larger organic molecules that assemble and react to form a complex systems chemistry sequence of reactions that would approximate a reverse citric acid (Krebs) cycle. Nick Lane provides more detail, especially about the vital role played by a proton gradient across thin FeS mineral "membranes" that facilitate carbon fixation, which requires the acceptance by CO_2 of electrons donated by H_2. In this reaction,

---

[5]An endergonnic chemical reaction consumes energy as opposed to energy releasing reactions which are exergonic.

CO_2 is said to be reduced (by accepting negatively charged electrons) and H_2 is said to be oxidized (by losing negatively charged electrons). The problem is that this reaction does not proceed at normal pH (proton concentration). Here is how Nick Lane describes the mechanism of this redox chemistry that he postulates provided the organic precursor molecules for abiogenesis (Lane N 2015):

> But now think of a proton gradient across a membrane. The proton concentration – the acidity – is different on opposite sides of the membrane. Exactly the same difference is found in alkaline vents. Alkaline hydrothermal fluids wend their way through the labyrinth of micropores. So, do mildly acidic ocean waters. In some places, there is a juxtaposition of fluids, with acidic ocean waters saturated with CO_2 separated from alkaline fluids rich in H_2, by a thin inorganic wall, containing semiconducting FeS minerals. The reduction potential of H_2 is lower in alkaline conditions: it desperately 'wants' to be rid of its electrons, so the left-over H^+ can pair up with the OH in the alkaline fluids to form water, oh so stable. . . . The only question is: how are electrons physically transferred from H_2 to CO_2? The answer is in the structure. FeS minerals in the thin inorganic dividing walls of microporous vents conduct electrons. . . . And so in theory, the physical structure of alkaline vents should drive the reduction of CO_2 by H_2 to form organics.

The question of how abiogenesis proceeds, once a sufficient supply of organic molecular precursors is concentrated within the micropores of alkaline hydrothermal vents, can be examined from different perspectives. A *constructionist* perspective approaches the question by trying to understand abiogenesis on the basis of first principles of physics and chemistry. A *deconstructionist* perspective asks what can be inferred about abiogenesis on the basis of the biology of contemporary life forms. The two approaches then can be used in conjunction in an effort to gain leverage on the difficult problem of the origin of life.

The Deconstructionist Approach: A Brief Outline of Molecular Biology

The central dogma of modern molecular biology involves the transfer of genetic information from DNA to RNA and then to proteins that are synthesized on the basis of genetic information carried by the RNA. The structure of Deoxyribonucleic acid (DNA) is illustrated in Fig. 5.1.

The polymeric DNA shown in Fig. 5.1 is folded into complex structures called chromosomes. The chromosomes of a human female are shown in Fig. 5.2.

The current view is that the human genome consists of approximately 20,000–25,000 protein-coding genes, which represents approximately 1.5% of all chromosomal DNA of the human genome. The vast majority of genomic DNA therefore consists of non-protein coding nucleotide sequences called *Introns*. Some *Introns* are regulatory regions that control selective expression of nearby genes at particular times during cell differentiation. The protein coding nucleotide sequences in genes are called *Exons*. *Introns* represent the greatest percentage of DNA and can be found both within protein-coding sequences (genes), as well as in long regions between genes. Their origin and potential significance is discussed below.

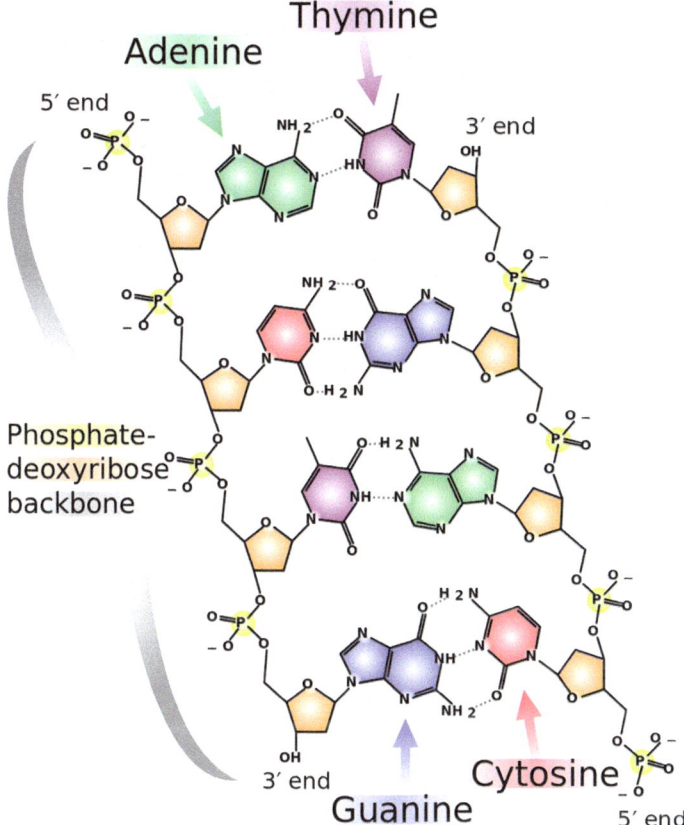

Fig. 5.1 DNA is a double-stranded polymer that is comprised of subunits. Each subunit, called a nucleoside, consists of a 2-deoxyribose sugar and one of four nucleobases: Guanine; Cytosine; Thiamine and Adenine. The deoxyribose sugar of each of these nucleosides binds to a phosphate group (PO⁻₄) to form what is called a nucleotide; and the nucleotides of each DNA strand are linked together by the phosphate groups to form the sugar-phosphate backbone of the polymer. Each nucleobase of one strand binds with a complementary base on the other strand in specific pairs: Guanine-Cytosine; Thiamine-Adenine. This file was made available under the Creative Commons CC0 1.0 Universal Public Domain Dedication. https://creativecommons.org/publicdomain/zero/1.0/deed.en

The synthesis of a new protein begins with the unfolding of a chromosome to expose a gene sequence contained in one region of the double helical structure of DNA. This is illustrated in Fig. 5.3.

The process of constructing a new protein molecule begins with transcription of the nucleotide sequence of DNA into the complementary nucleotide sequence of messenger RNA (mRNA) in the cell nucleus. A schematic of this process is shown in Fig. 5.4.

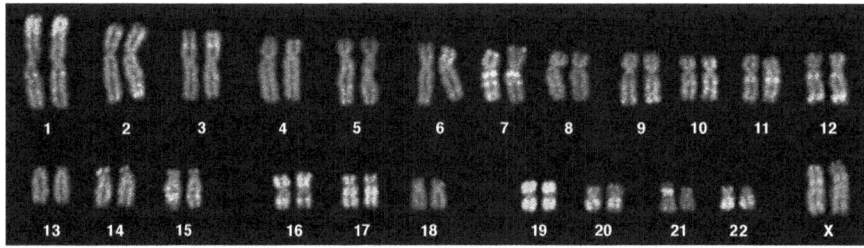

Fig. 5.2 Karyotype from a female human lymphocyte (Bolzer et al. 2005). The public domain image was downloaded from: https://upload.wikimedia.org/wikipedia/commons/2/27/PLoSBiol3. 5.Fig7ChromosomesAluFish.jpg and is reproduced here under the Creative Commons License found at: https://creativecommons.org/licenses/by/2.5/deed.en

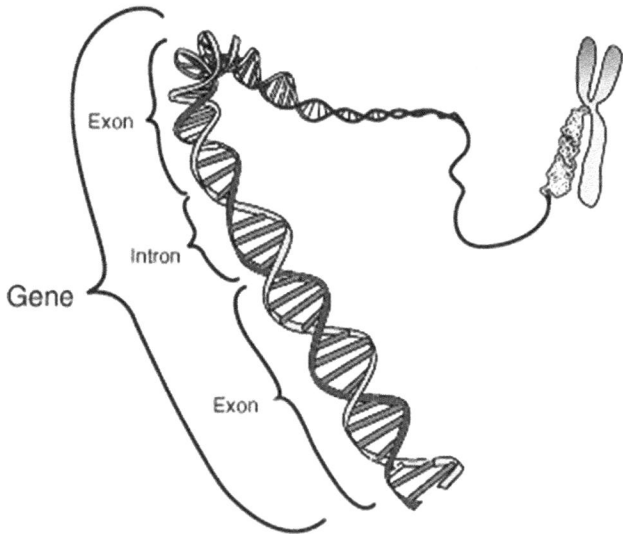

Fig. 5.3 Unfolding of a chromosome to reveal a gene that contains two exons on either side of an intron. The gene shown contains an intron and two exons. The exons are the protein coding regions. This public domain image was downloaded from: http://upload.wikimedia.org/wikipedia/com mons/0/07/Gene.png

Once transcribed, the mRNA must be processed to remove the *intron* segments before translation of the message begins at the ribosome. This is accomplished in the cell nucleus in the process illustrated in the schematic of Fig. 5.5.

The mRNA then is trafficked (transferred) out of the cell nucleus to the cytoplasm where its message of triplet nucleotide sequences is translated at a special organelle called the ribosome. The ribosome is itself composed of another type of RNA called ribosomal RNA. The message that is carried by mRNA consists of triplet nucleotide

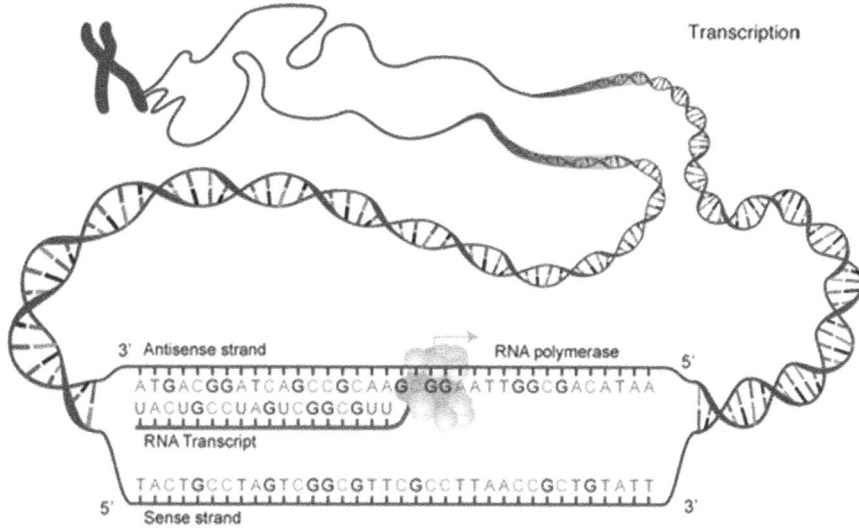

Fig. 5.4 Transcription ("writing") of the genetic "message" encoded in the nucleotide sequence of DNA to a complementary strand of RNA. The RNA strand created by this process carries the genetic message to the protein synthetic machinery outside the cell nucleus. Because this form of RNA carries the genetic message it is called messenger RNA (mRNA). RNA is the acronym for ribonucleic acid. Unlike double-stranded DNA, RNA is single-stranded and uses a ribose sugar instead of a deoxyribose as part of its sugar-phosphate backbone. The shaded region represents the enzyme, RNA polymerase, the enzyme that polymerizes RNA from nucleotide subunits. RNA polymerase first must unwind the double stranded DNA helical structure to expose a region that is to be transcribed. The RNA polymerase next adds a corresponding nucleobase to the emerging mRNA strand, and then rewinds the two DNA strands. This public domain image is from the US National Library of Medicine website at: https://geneed.nlm.nih.gov/images/transcription_sm.jpg under terms specified at: https://www.nlm.nih.gov/copyright.html

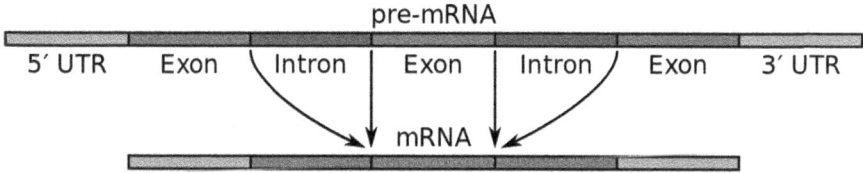

Fig. 5.5 Schematic illustration of an un-spliced mRNA precursor, with two introns and three exons (top). UTR = untranslated region. After the introns have been removed via splicing, the mature mRNA sequence (bottom) is ready for translation outside the nucleus. This public domain image was downloaded from: https://commons.wikimedia.org/w/index.php?curid=7063375

sequences, referred to as codons, that each specify a particular amino acid that is to be added to the growing polymer of amino acids in the sequence specified by the mRNA to form the protein encoded by the gene. Amino acids are carried to the ribosome by yet another type of RNA, known as transfer RNA or tRNA, as shown in Fig. 5.6.

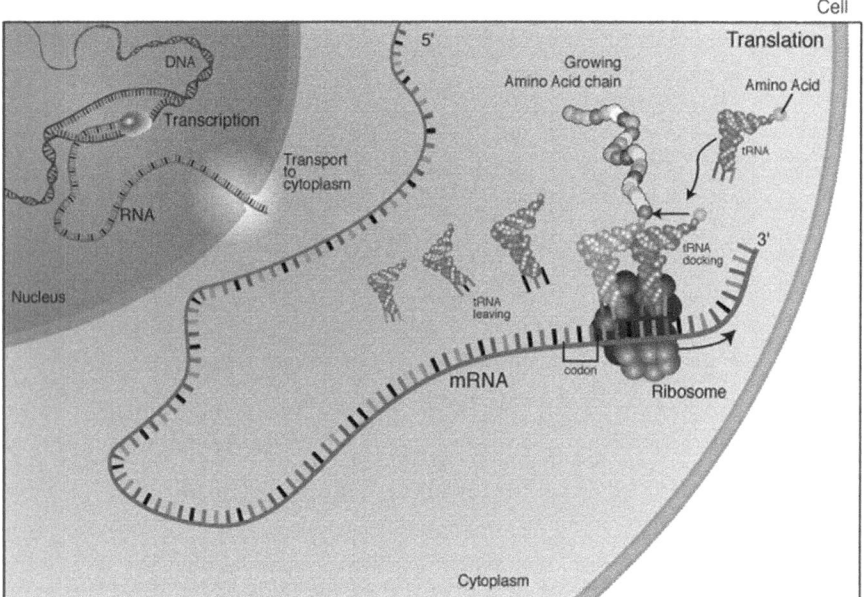

Fig. 5.6 Each amino acid is recognized and bound by a cognate tRNA that only binds to that amino acid. The tRNA has a triplet nucleotide sequence, called the anti-codon, that matches and binds to the corresponding codon on the mRNA in the ribosome. The polypeptide chain grows as tRNA molecules bring the appropriate amino acids in a sequence that is ultimately specified by the nucleotide sequence of the gene that codes for that particular protein. This public domain image was downloaded from the US National Institutes of Health at https://geneed.nlm.nih.gov/images/translation_lg.jpg under terms specified at https://www.nlm.nih.gov/copyright.html

The growth of the amino acid chain is accomplished by the formation of a peptide bond between each new amino acid added to the chain and the one added prior to it. The reaction is shown in Fig. 5.7. One amino acid loses a hydrogen atom and an oxygen atom from its carboxyl group (COOH), while the other loses a hydrogen atom from its amino group (NH_2). The two released hydrogen atoms bind to the oxygen released by the reaction to form one molecule of water (H_2O), while the two amino acids are joined by a peptide bond (-CO-NH-).

The information presented above represents the empirically established mechanism for the transfer of information stored in genomic DNA to mRNA, and how it is used at the ribosome in conjunction with tRNA to determine the specific amino acid sequence of each protein. Virtually all of these steps are regulated and catalyzed by proteins in cells today. The interested reader can find the definitive history of the revolution in molecular biology portrayed in H. Freeland Judson's wonderful book "The Eighth Day of Creation" (Ibid). This book deserves special mention here as one

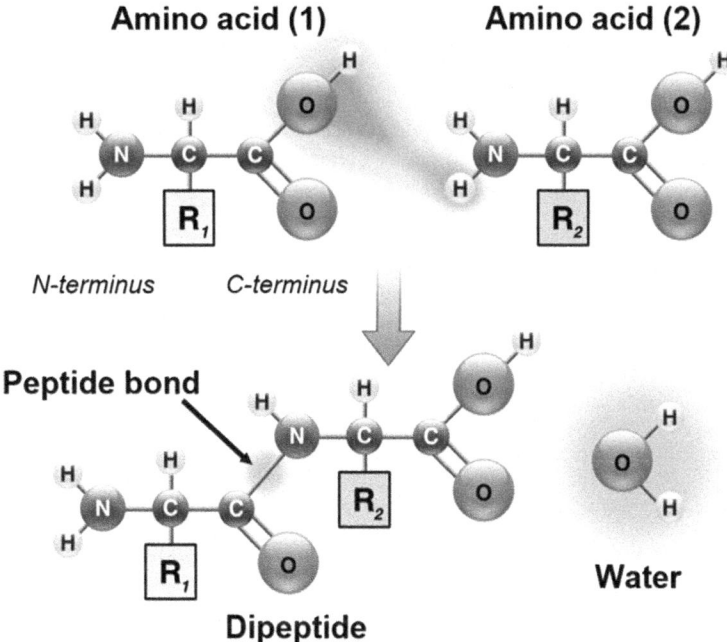

Fig. 5.7 The reaction mechanism for the formation of a peptide bond as explained in the text. This public domain image was downloaded from: https://upload.wikimedia.org/wikipedia/commons/6/6d/Peptidformationball.svg and is used under terms specified at: https://commons.wikimedia.org/wiki/File:Peptidformationball.svg

of the great works of science history. Judson presents the definitive account of one of the most important and disruptive epochs in the history of science. He imbues the scientific story with biographical and conceptual perspectives that provide a vivid sense of the drama in the race to discover "the secret of life", as the mechanism of inheritance was then called. This insightful book sheds light on how researchers in the nascent field of molecular biology made the transition from uncertainty to insight to solve the many questions posed by humanity's desire and quest to understand life.

Less clear is how life and its molecular biological mechanisms arose on the primitive Earth from non-living matter. We have no satisfactory theory for the relative timings of the appearance of RNA and proteins in abiogenesis. Nor do we have a satisfactory understanding of how the genetic code was established by which the sequence of RNA nucleotide triplets determines the sequence of amino acids in the protein encoded by the RNA. It may be helpful, therefore, to consider the potential interactions among amino acids, polypeptides and RNA molecules during the epoch of chemical evolution that led to the modern state of affairs.

The Constructionist Approach: Building Life from the Ground Up

J. Desmond Bernal was the first to suggest a critical role for clay in the earliest stages of chemical evolution that led to biogenesis (Bernal J D 1951). Commenting on this, Kavita Gururani and co-authors state that (Gururani K Pant CK and Pathak HD 2012):

> Bernal suggested that clays might have played a significant role in the primitive earth (sic) through the process of concentration and adsorption of the biologically formed biomonomers and thus protecting them against hydrolytic fission...
>
> It is assumed that clay minerals and metal oxides near sea shores or sea beds may have played a key role in concentration of biomonomers through adsorption and desorption processes on their surfaces. The importance of clay minerals in chemical evolution was first suggested by Bernal in 1951. [Bernal] proposed that clays near the hydrosphere-lithosphere interface might have adsorbed biomonomers on and between their silicate layers and then facilitate[ed] condensation considerably leading to the formation of biopolymers and protecting them from hydrolysis.

Gururani and colleagues (Ibid) went on to demonstrate experimentally that the amino acid adenine readily adsorbed to Montmorillonite clay, and that this adsorption was enhanced when divalent cations such as Ca^{2+}, Mg^{2+}, or Cu^{2+}, were incorporated into Montmorillonite as Ca^{2+}-Montmorillonite, Mg^{2+}-Montmorillonite, or Cu^{2+}-Montmorillonite.

A review of the role of clay catalysis of organic molecule synthesis, and even amino acid polymerization into polypeptide structures in primitive earth environments states (Hashizume H. 2012):

> Clay minerals would be capable of adsorbing bio-organic molecules from the early ocean. The resultant clay-organic complexes would partly be deposited on the ocean floor. Greenland and colleagues investigated the interactions of various amino acids with H-, Na-, and Ca-montmorillonites [References Omitted]. Arginine, histidine, and lysine adsorbed to Na- and Ca-montmorillonites by cation exchange. Other amino acids (alanine, serine, leucine, aspartic acid, glutamic acid, phenylalanine) adsorbed to H-montmorillonite by proton transfer. The adsorption of glycine and its oligo-peptides by Ca-montmorillonite and Caillite increased with the degree of oligomerization (extent of polymerization and the resulting molecular weight). Hedges and Hare (1987) suggested that the amino and carboxyl groups of the amino acids were involved in their adsorption to kaolinite.

Amino acids that formed in the warm environment of alkaline hydrothermal vents, as postulated by Lane (Ibid) and others, could have adsorbed to clay on vent surfaces, or on suspended clay particles in and near the vents. A reaction mechanism that is consistent with the role of carboxyl and amino groups in the adsorption of amino acids to clay that was suggested by Hedges and Hare may have involved the chelation of metal components of clay by those same groups. It is known for example that such metal chelates can be synthesized in a chemical reaction between the dipeptide amino acid, L-carnosine, and transition metals such as zinc, iron, strontium and others (Baran EJ 2000; Matsukura T and Tanaka H 2000). The reaction mechanism yields the amino acid-metal chelate, L-carnosine zinc. In this chemical structure, the amino and carboxyl parts (moieties) of the molecule bind to zinc to form the chelate, in which the zinc is coordinated to two nitrogen atoms and one oxygen atom in L-carnosine.

Hashizume (Ibid) also noted that the catalytic property of clay could facilitate the formation of nucleotide polymers of varying lengths, as well as the formation of peptide bonds between adjacent amino acids adsorbed by various mechanisms to a clay surface, or between free floating clay-amino acid complexes. Polymerization of clay-activated amino acids via peptide bond formation would produce polypeptides of various lengths which would then assume three-dimensional configurations based on hydrophobic[6] and hydrophilic[7] interactions with water once released from the clay in the aqueous environment of the primitive seas. Some of these peptidic-proteinaceous structures might have had conformational and energetic properties that are consistent with catalytic activity that would energetically favor and acceler-ate the synthesis of other organic molecules and polymers such as ribonucleic acids (RNA).

Following upon the work of Bernal A., Graham Cairns-Smith proposed a mech-anism for the transition from inorganic chemistry to organic carbon-based chemistry necessary for life (Cairns-Smith A G 1982). Starting with the assumption that the first self-replicating molecules would be the simplest to synthesize, they were most likely to have been those that could be formed on the principles of inorganic chemistry. On this basis, Cairns-Smith proposed that inorganic clay polymers provided not only an activating substrate for organic molecule adsorption, but also at least some of the information required for the synthesis of the first self-replicating polynucleotides. Clay is a crystalline mineral, in which the atomic constituents form repetitive (polymeric) structures that comprise a pattern for successive layers of clay that form one upon the other. Each layer provides the template for the formation of the next layer. The pattern of atomic constituents therefore represents a kind of inorganic "proto-genetic code" *for the crystallization (replication) of clay* that Cairns-Smith proposed was "taken over" by organic molecules that would bind to specific regions of the clay surface under the right conditions. The resulting pattern of organic molecules could have become polymeric ensembles in the event that the organic monomers could form inter-molecular bonds before or during separation from the clay. Although its antecedents were present, the genetic code as we understand it would not have existed at this stage. RNA nucleotide sequences formed on the basis of clay activation-catalysis would have had a random character; or at best these primordial nucleotide sequences would have been loosely determined by modest affinities of particular nucleotides to specific regions on the clay surface. The information content in the sequence of nucleotides in this type of primordial RNA would have been no greater than the information contained in configuration of adjacent regions on the surface of clay. Similarly, the sequence of amino acids in peptides formed by the kind of non-guided polymerization afforded on the basis of

---

[6]Hydrophobic means that molecular interaction with water molecules is not energetically favored. Hydrophobic substances are insoluble in water, therefore.

[7]Hydrophilic means that molecular interaction with water molecules is energetically favored. Hydrophilic substances are therefore water soluble.

clay activation-catalysis described above would have had a random character. How then did this state of affairs lead to the emergence of the modern genetic code? It is estimated that approximately 600,000,000 years passed between the time the Earth was formed and life first appeared on the planet. So, there was sufficient time, water and energy for inorganic chemistry to transition to organic chemistry through the type of geochemistry found in alkaline hydrothermal vents. Conditions in alkaline hydrothermal vents were consistent with the production of amino acids, as well as clay-mediated synthesis of peptides of various lengths and conformations. In addition, there are good grounds for believing that the alkaline hydrothermal vent environments of the primordial seas were capable of supporting nucleotide synthesis and polymerization to produce RNA polymers of various lengths. The origin of the genetic code must be understood, therefore, in terms of the interactions among the earliest polynucleotides, peptides and amino acids. Further pursuit of this and related questions in the field of abiogenesis is beyond the scope of this book, but a deeper exploration of abiogenesis is provided by Nick Lane's Book, "The Vital Question: Energy, evolution and the Origins of Complex Life" (ibid). Among its many virtues is that Lane disdains the debate concerning metabolism first versus genes first, and embraces the necessity of co-evolution of energy metabolism and the molecular mechanisms of inheritance.

References

Baggott J (2015) Origins. Oxford University Press, New York, pp 215–222

Baran EJ (2000) Metal Complexes of Carnosine. Biochmistry (Moscow), 65(7):789–797. Translated from Biokhimiya 65(7):928–937 (2000)

Bernal JD (1951) The physical basis of life. Routledge and Kegan Paul, London

Bolzer A, Kreth G, Solovei I, Koehler D, Saracoglu K, Fauth C, Müller S, Eils R, Cremer C, Speicher MR, Cremer T (2005) Three-dimensional maps of all chromosomes in human male fibroblast nuclei and prometaphase rosettes. PLoS Biol 3(5):e157. https://doi.org/10.1371/journal.pbio.0030157

Cairns-Smith AG (1982) Genetic takeover and the mineral origins of life. Cambridge University Press, New York

Gururani K, Pant CK, Pthak HD (2012) Surface Interactino of adenine on Montmorillonite clay in presence and Abasence of divalent Cations in relevance to chemical evolution. Int J of Scientivic and Tech Res 1(9):106–109

Hashizume H (2012). Role of clay minerals in chemical evolution and the origins of life, clay minerals in nature – their characterization, modification and application, Dr. Marta Valaskova (Ed.), InTech, https://doi.org/10.5772/50172. Available from: http://www.intechopen.com/books/clay-minerals-in-nature-their-characterization-modification-and-application/role-of-clay-minerals-in-chemical-evolution-and-the-origin-of-life

Hedges JI, Hare PE (1987) Amino acid adsorption by clay minerals in distilled water. Geochim Cosmochim Acta 51:255–259

Lane N (2015) The vital question. WW Norton and Company, New York, pp 115–121

Matsukura T, Tanaka H (2000) Applicability of zinc complex of L-carnsoine for medical use. Biochem Mosc 65:817–823

Chapter 6
Paleopsychology: The Emergence of Mind in the Universe

Richard J. Di Rocco and Edgar E. Coons

> *Every block of stone has a statue inside it and it is the task of the sculptor to discover it.*
>
> Michelangelo

> *Most of malaria's victims lie in a belt across the middle of the Earth. . . . [they] share with us this moment in time, in the continued evolution and sculpting of the human genome – and the human genome takes notice.*
>
> Caporale (2003)

> *My strength is made perfect in weakness.*
>
> 2 Corinthians 12:9

Abstract The vital role of *entropy and information* in the origin of life from non-living matter, and the evolution of living matter once it was established on Earth, are examined in this chapter. The origin of human and animal learning and memory in the signal transduction mechanisms of pre-Cambrian single-celled organisms of the primordial oceans is described. This phenomenon offers one of the strongest examples of conserved cell and molecular mechanisms in biology. The subsequent rise of multicellular organisms during the Cambrian explosion approximately 500 million years ago is discussed in the context of predator-prey relationships among the invertebrate animals that were living at that time. The predator-prey

R. J. Di Rocco (✉)
Psychology Department, University of Pennsylvania, Philadelphia, PA, USA

Psychology Department, St. Joseph's University, Philadelphia, PA, USA
e-mail: richdi@upenn.edu

E. E. Coons
Psychology Department, New York University, New York, NY, USA
e-mail: eec1@nyu.edu

© Springer Nature Switzerland AG 2018
R. J. Di Rocco, *Consilience, Truth and the Mind of God*,
https://doi.org/10.1007/978-3-030-01869-6_6

dynamic provided the intense selective pressure for the evolution of neural networks that were optimized for effective escape and predatory behaviors, as well as for the mechanisms of synaptic plasticity that underlie learning and memory. The rise of the vertebrates, and mechanisms of learning in birds and mammals, is described as the prelude to the emergence of the modern human mind with its amazing cognitive fluidity that gave rise to creativity, the power of abstraction and symbolic thought. The dawn of human meta-awareness in *Homo sapiens sapiens*, the *human who knows he knows*, is described and its implications for the emergence of existential fear is discussed. One of the consequences of increased intelligence, besides its ability to magnify fear of the unknown, is the corresponding compulsion to provide explanations for phenomena that are not understood. This characteristic of modern human thinking has been referred to as the cognitive imperative, and it is manifested in the magical thinking of early *Homo sapiens*, as well as contemporary children. Existential dread, the fear of annihilation of the ego at death, is discussed as the likely driving force for the emergence of mythology as proto-religious dogma that provided a sense of comfort and reduction of anxiety in our human ancestors, as well as in contemporary humans. A theory of the origin of evil in human behavior is offered, in which the overvaluation by humans of the hypothetical constructs that we generate to explain the unknown provides a powerful impetus for genocide and global war. We are the only animals that kill members of our own species to defend ideas, because we fear the loss of the security that our constructs of reality provide.

Keywords Life and information · Signal transduction · Dendritic spines and synaptic plasticity · Sign-tracking in mammals and birds · Evolution of cytoarchitacture in prefrontal cortex · Modern human mind

The Pervasive Influence of Entropy on Biological Evolution

We considered in the last chapter some potential scenarios for abiogenesis that involve various processes of chemical evolution in geochemical environments of the early Earth. The first great achievement of abiogenesis was the establishment of organic polymers, most likely short to medium length peptides and polymeric nucleotides such as RNA that were capable of self-replication. Modern science fully embraces the idea that it is possible to explain this remarkably improbable event entirely on the basis of the laws of physics and chemistry. Nevertheless, while the establishment of self-replicating molecules is a necessary stepping-stone on the path to life it is not sufficient. Something more was required. That something is the aspect of metabolism called bioenergetics, the mechanisms that capture order from the unidirectional[1] photons emanating from the sun and convert that order into high-energy phosphate bonds of the energy currency of life known as adenosine triphosphate (ATP). ATP contains three phosphate groups bound into the molecule's

---

[1]Unidirectional because the photons emanate from the sun, which is a point-source of light.

structure by high-energy chemical bonds. Hydrolysis, or breaking, of one of the high-energy phosphate bonds releases energy that can be captured and used to perform various kinds of biological work. This is the work that must be done to maintain the highly-ordered state of living systems, which is far removed from thermodynamic equilibrium. Metabolism is life's defensive adaptation against the disordering of macromolecular structure by entropy.

Once self-replicating molecules became established on Earth, their modification according to the laws of physics and chemistry would have become determinants of the early evolution of life. What physical laws or forces would have been relevant to the evolution of life? We can triangulate a bit here using what biology tells us about evolution to ask the question another way. What physical factor constitutes the essential quality of what Darwin called *selective pressure,* a key determinant in the natural selection of adaptive traits that enhance survival? Clearly *selective pressure* assumes diverse forms, but careful consideration leads to the hypothesis that a tendency to increase the entropy of the organism is the essential physical property possessed by all of them. Living beings maintain the matter of which they are composed in an extremely organized state. For any open system, like an organism that can exchange matter and energy with the surrounding environment, the Second Law of Thermodynamics states that there is an inexorable tendency for the system to become more disordered and to move toward equilibrium with its surroundings over time unless work is done to maintain the highly ordered molecular structures that are necessary for life to endure. Thermodynamic equilibrium is achieved for an organism when its entropy is at a maximum, and this occurs with death and decomposition. When a living being comes into thermodynamic equilibrium with its surrounding it is dead. This is exemplified by the cooling of a dead body and its ultimate decay into the constituent atoms of which it is composed. With respect to its impact on life, therefore, entropy may be viewed as the most fundamental source of what Darwin referred to as "selective pressure". *All forms of "selective pressure" are commonly viewed as life threatening in varying degrees and may be understood as agents of entropy.*

This idea leads to the thesis that there is an *entropic imperative* for the emergence and evolution of life and mind. That is, given sufficient time, space, matter, and energy, entropy selects for the molecular structures and other adaptations that are necessary for the emergence of life and its evolution to increasingly better adapted forms. Random genetic mutations, that code for adaptive anatomical, biochemical or physiological traits, confer an advantage by helping the organism resist the effects of entropy. Adaptive traits help the organism to survive to the point of successful reproduction and are therefore passed on to succeeding generations. This is natural selection as defined in Darwinian Biology; and it provides the essential mechanism of evolution. Paradoxically, despite its life-threatening effects, we see that entropy provides the universal basis for selecting from among the random genetic mutations and associated adaptations the ones that offer the best defense against the adverse effects of entropy. It is remarkable that the destructive and disorganizing action of entropy provides the means by which protective mechanisms that defend life against entropy's deleterious effects are selected for transmission to succeeding generations.

We will see below that the mechanisms of bioenergetics, neural networks, brains, and mind, in particular, may all be viewed as key adaptations that provide selective advantages in the struggle against the life-threatening effects of entropy.

What About the Cockroach?

The failure of the cockroach to evolve into a more intelligent life form over hundreds of millions of years can be invoked to refute the hypothesis of an *entropic imperative* in the evolution of increasingly complex and intelligent forms of life. That is, if entropy impels evolution toward ever more complex and intelligent forms of life, why isn't the cockroach more intelligent? To answer this question, we must appreciate that entropy does not vary at a steady rate in relatively small regions of the universe like the Earth. There are times when the environment remains relatively stable and other times when more dramatic upheavals occur. The potential impact on evolution of this uneven variation of entropy was captured by paleontologists Stephen Jay Gould and Niles Eldredge in their theory of punctuated equilibria (Gould and Eldredge 1977).

The observation of long periods of relative stasis in the fossil records of most species followed by the "sudden" appearance of dramatic changes, led Gould and Eldredge to realize that the gradual evolutionary change envisioned by Darwin may not have been sufficient to explain the emergence of new species, an event known as *speciation*. This seems reasonable because during periods of relatively stable entropy production by the environment an organism's biological niche does not change sufficiently to threaten its survival. In this case, the magnitude of entropy change does not provide sufficient selective pressure on genetic variations for a speciation event to occur. Alternatively, the change in entropy may be sudden and large enough to severely challenge the organism's adaptive survival mechanisms. Under these conditions only those individuals from among existing genetic variants that possess adaptations sufficient for survival will endure, while other variants do not. A change will be accomplished that manifests as a relatively sudden expansion of the population that possesses the successful genetic variant. If the environmental shift endures long enough, or the successful variant becomes isolated for multiple generations, genetic changes will accumulate in a relatively short evolutionary timeframe. A new species would emerge that is well-adapted to the new ecological circumstance. The new species would then account for an ensuing long period of evolutionary stasis until the entropy change again exceeds the species' adaptive mechanisms and another sudden shift would occur in the paleontological record.

The existence of long periods of evolutionary stasis does not refute the importance of an entropic imperative in evolution. Rather it demonstrates the adequacy of a species' adaptations in the face of all of the environmental changes that have occurred since it first emerged. So, the record of protracted evolutionary stasis of the cockroach may be understood simply in terms of the enduring adequacy of its adaptations during periods of dramatic environmental change that other species were less able to handle. The evidence in favor of an entropic imperative is found in the sudden emergence of new organisms when the magnitude of entropy change severely challenges the capacity of existing adaptations of a species. This is captured in the aphorism related to the evolutionary descent of birds from dinosaurs. You may have heard or read this as: "The dinosaurs didn't become extinct. They just flew away". Finally, we should observe that the thesis of an entropic imperative in the evolution of life and mind does not state that every species must evolve to a super intelligent state. Rather the thesis states that, given enough time, entropy tends to select for better adaptations among which intelligence is one that is bound to emerge in at least *some* species. Accordingly, an excellent counter-example to the Cockroach argument against the entropic imperative is provided by the Octopus, a member of the invertebrate phylum, Mollusca.[2] This remarkably intelligent neurologically advanced animal is a member of the largest invertebrate phylum, the Mollusks, and has a nervous system and behavior that approaches that of many mammals in complexity and sophistication.

Another question that often arises in discussions of the relationship between entropy and evolution concerns why the growth of order that is essential to life does not violate the Second Law of Thermodynamics, which requires that the disorder of the universe always increases. The answer involves the distinction between local and universal entropy. It is clear that the maintenance of life requires preservation of order in the unique arrangement of molecules that is consistent with the physiological and metabolic processes of the living state. In this case, the Second Law requires that the local increase in order occurs at the expense of a larger increase in entropy for the universe as a whole. This is precisely what happens when, owing to gravitational clustering, stars and planetary systems form from galactic gas and dust. The local growth of order is what allows life to emerge in planetary breeding grounds, while the required local reduction of entropy is offset by a greater increase in entropy for the universe as a whole.

Life and Information

The solar system, and the life it has given rise to on the Earth, raises the prospect that other stars also may have planets that provide sufficient breeding grounds for life. We now know that there are many other stars with planetary systems, even ones that possess planets like Earth. Certainly, it is reasonable to entertain the hypothesis that

---

[2]The *Mollusca* includes four classes of animals that can be found in fresh or salt water, and land: Gastropods (snails); Bivalvia (clams, scallops, mussels); and Cephalopods (octopuses, squid and Cuttlefish), and finally Polyplcophora (chitons). Many of these species are edible and are harvested along coastal regions or from lakes, streams and land. Saltwater Mollusks are thought to have provided a major source of food to *H. sapiens* as our ancestors migrated out of Africa along the coastal route of East Africa to the Arabian Peninsula and the Levant before moving into Europe and Asia.

life arose more than once in this vast universe of billions of galaxies and trillions of stars. From this point forward, however, we will be concerned with the evolution of single-celled and multicellular organisms on Earth. Sometime after the establishment of self-replicating biopolymers the metabolic and genetic machinery, that is essential for the maintenance and propagation of the ordered state we call Life, was packaged into cells. Considerable uncertainty persists about how this happened, but again there are reasonable ideas that are consistent with physical laws that have been brought to bear on the question. The interested reader can pursue this issue in the cited works of Nick Lane (Ibid) and Jim Baggott (Ibid).

In his seminal book, "What is Life", Erwin Schrödinger postulated that to avoid the rapid decay to thermodynamic equilibrium known as death, *living organisms recognize, approach and assimilate order* as the antidote to the life-destroying effects of entropy (Schrödinger 1944).[3] Where does this order come from? How is it assimilated? As pointed out by Roger Penrose, the sun is the local source of order in the solar system because the radiant energy of the sun that hits the Earth derives from a single source in the sky (Penrose 2011). These photons that reach us from the sun have shorter wave-lengths (higher frequency) and are therefore more energetic than the photons in the cosmic background radiation. The radiation that reaches Earth from the sun is ordered because it comes from a point source that stands out against the high wavelength radiation of the cosmic background. This allows the energy it carries to be efficiently captured by photosynthetic processes in plants via the energy transfer that is utilized to create complex carbohydrate molecules. The carbohydrate molecules contain more order than the carbon dioxide and water molecules from which they are synthesized. *The ordered nature of light from the sun is captured into the highly-ordered structure of complex carbohydrate molecules via the agency of energy transformations of the plant's metabolism.* The order and energy that is stored in the chemical bonds between atoms of carbohydrate molecules is captured in turn when these molecules are catabolically broken down in the plants themselves, or the animals that eat them. Once again, energy acts as the agency to transfer order in anabolic biosynthetic activity to create the high-energy phosphate bonds of ATP, when ADP reacts with inorganic phosphate. ATP provides the requisite energy for the performance of diverse forms of biological work that are needed to protect and maintain the highly-ordered structure of the organism. Examples of biological work include not only the aforementioned anabolic processes of biosynthesis, but also muscle contraction, neuron firing and active transport of molecules and ions across cell membranes. In this manner, a large part of the order that is extracted from the sun's radiance is acquired by the food chain to maintain the ordered structure necessary for the maintenance of life. In this context, it is important also to note Penrose's (Ibid) essential point that the total amount of

---

[3]The impact of Schrödinger's book was praised by H. Freeland Judson in *The Eighth Day of Creation*, which remains the definitive history of the revolution in molecular biology. Judson observed that Schrödinger provided the motivation for many young physics students to bring their skills and methods to the study of key questions concerning the nature of the hereditary material, as well as the structure and synthesis of proteins.

energy that the Earth radiates back into space is equal to the amount of energy absorbed from the sun, but the entropy of this radiation is much higher. It consists of higher wavelength (lower frequency, lower energy) photons than the photons from the sun. What is extracted from the sun's radiance, therefore, is the order required to maintain the living state despite the degrading effects of entropy! The order extracted from the sun's radiance accounts for the corresponding amount of disorder, or entropy, in the radiation that is radiated back into space by the Earth.

Antecedents of Mind

More remarkable than the bioenergetic consequences of assimilating order and storing it in the form of complex molecules, such as carbohydrates, fats and proteins, is the ability of organisms to detect and approach order in the form of information in the environment. The phototropism of plants is perhaps the best-known example of this and an excellent validation of Penrose's and Schrödinger's thesis about the sun being the source of order for the organisms of Earth. Even early Prokaryotes, the simplest forms of single-celled organisms such as the bacteria, were capable then as they are now of recognizing and approaching order while avoiding entropy. This has been amply demonstrated by the ability of the bacteria to approach food and avoid noxious stimuli. (Parkinson 1993). The same capabilities have been observed in more advanced Eukaryotic single-celled organisms called Protozoans, such as the Amoeba and Paramecium (Maier and Schneirla 1935). The mechanisms of signal transduction and intracellular molecular signaling that are responsible for the ability of single-celled organisms to behave in this manner have been preserved in the cells of multi-cellular (metazoan) animals and have been shown to substantially account for the molecular mechanisms of learning and memory in vertebrates including humans. All cells have mechanisms for detecting extracellular stimuli present at the cell membrane surface. Interaction of the stimulus molecule with a membrane receptor triggers phosphorylation of intracellular proteins known as kinases, which phosphorylate other kinases in a cascade that ends with phosphorylation of a transcription factor.[4] The transcription factor is a protein that binds to a regulatory region of a gene to initiate DNA transcription that produces mRNA. In turn, the mRNA is used to synthesize new proteins that are required by the cell to optimize the cellular response to the extracellular stimulus in the future. The overall process involves signal transduction at the membrane and intracellular signaling via the kinase phosphorylation cascade, and ultimately results in new protein synthesis (Kyriakis and Avruch 2001). While this mechanism is ancient phylogenetically, it is universally present in metazoans, including the cells of the human body. In fact, this is the mechanism used by neurons to bring about the cellular changes that are responsible for learning and memory, as demonstrated by Eric Kandel in his Nobel

---

[4]Phosphorylation adds a phosphate group (PO_43-) to another molecule.

Prize winning work on the molecular mechanisms of learning in the Sea Snail, Aplysia (Kandel 2000).

During the so-called Cambrian explosion, approximately 542 million years ago, a veritable "Big Bang of Biology" occurred in which many forms of Metazoan life made their first appearance on Earth (Koonin 2007). These animals were invertebrates. While they lacked a backbone, many had nervous systems ranging in complexity from the simplest of neural nets to more complicated nervous systems of the primitive Cephalopods that consisted of many neurons. The simplest group among the surviving invertebrates, which is known as the Porifera, or sponges, lacks neurons completely. The adult organisms exist in a sessile state attached to the sea floor. Porifera therefore have a reduced need to coordinate complex activity among cells. The degree of integrative activity that they require is mediated by ion channels in certain groups of adjacent cells that are thought to be potential precursors of neurons that are found in other species. Simple nervous systems allowed motile metazoans[5] to solve the problem of coordinating the action of many cells. It is in these creatures that the ability to recognize and approach order in the environment was taken to a higher level as an adaptation to the selective pressure imposed by predator-prey relationships. Survival of motile metazoans depended on the capacity for complex movements in the three-dimensional space of the primitive oceans. Coordinated complex movement necessarily required neural representations of *self*-position and *other*-position in space to effect adaptive predatory or escape movements. To be effective, such neural representations would require continuous updating in real-time, thus forming an internal representation of *self in four-dimensional* space-time. In its simplest manifestation, this representation takes the form of *self-toward prey* and *self-away from predator*.[6] A clear advantage would result from larger and more complicated nervous systems with greater computational power. More sophisticated nervous systems would be capable of a greater degree of sensory information processing and would have the ability to compute and execute more complex movement strategies that would lead to greater success in the predator-prey competition. From this, it can be seen that the intense selective pressure of predator-prey relationships during the Cambrian epoch provided an impetus for the emergence of increasingly complex neural networks, sensory-motor integration mechanisms and correspondingly complex behavioral repertoires.

At all levels of neural processing, from receptors on the body surface to neural networks deep within the central nervous system, various subsets of neurons interact to form representations of discrete perceptual elements that represent features of stimuli present in the environment. Certainly, the primitive nervous systems of Cambrian metazoans would likely have had a keen sensitivity to those stimulus features that were associated with predators or prey. In this manner, biologically

---

[5]The ones not fixed to one location by attachment to a substrate, i.e. the ones that have locomotion.

[6]This is seen in the dead reckoning exemplified by a sand crab that meanders along a tortuous path on the beach, but nevertheless makes a "bee line" back to the safety of its burrow at the first sign of danger.

significant objects were located in the neural representation of space-time; and neural networks were configured to detect changes in their position for the obvious reason that movement might indicate an approaching predator or an appetizing prey species. We can see from this that the detection of changing stimulus intensity and position was a vital task carried out by the earliest nervous systems. Another critical task and capability of these earliest nervous systems was even more important, however.

Even the primitive nervous systems of many invertebrates are configured to detect information that manifests as the spatial and temporal correlations that occur between biologically relevant stimuli and other stimuli of lesser significance. By virtue of such correlations or associations the neutral stimuli acquire the ability to convey information about the temporal immanence or spatial proximity of biologically significant stimuli. In the mammals, for example, the proverbial snapping twig may warn the deer drinking at the water's edge that a tiger is about to pounce, or the presence of vultures in the sky above may indicate to the hyenas that a carcass is available for the taking. Organisms must pay attention to correlations between otherwise neutral stimuli and biologically significant ones because correlation provides vital information. For this reason, one may argue that the ability to detect and respond appropriately to the information that is inherent in correlated stimuli is the most important thing that nervous systems do. The information that is contained in correlated stimuli is recognized initially in a learning process known as Pavlovian, or Classical, Conditioning. Once the relationship of a neutral stimulus as a predictor of a biologically significant one has been learned, however, the organism will benefit from the predictive value of the neutral stimulus as an advance warning of danger or opportunity. Judging from its widespread occurrence among extant invertebrate phyla, it is likely to have made its first appearance in the Cambrian. The early phylogenetic appearance of Classical Conditioning in the invertebrate Metazoans provides a clear indication of the importance of detecting information (order) as a countermeasure to entropy and is consistent with Schrödinger's ideas about the significance of information for the survival of organisms. We obviously cannot experimentally demonstrate Classical Conditioning in any of the invertebrates of the Cambrian. Many contemporary invetebrate species have remained relatively stable since the Cambrian epoch (much as the Cockroach) and Pavlovian conditioning has been demonstrated in many of the living descendants of the Cambrian invertebrate species. So, it is reasonable to assume that their Cambrian ancestors shared this ability.

Learning in Mammals and Birds

Classical conditioning was first demonstrated by the Russian physiologist Ivan Pavlov. Pavlov was studying the salivation response in dogs by measuring the saliva elicited by the presentation of meat powder in the dog's mouth. He noted that when the presentation of meat powder was preceded regularly by a sound, such as a ringing bell, the bell eventually acquired the capacity to elicit salivation

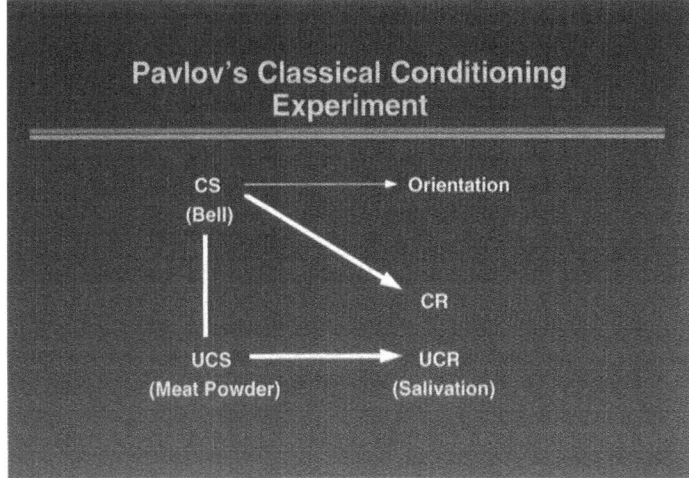

Fig. 6.1 Illustration that shows the pairing between a biologically relevant stimulus, the Unconditional Stimulus (UCS), and a neutral stimulus called the conditional stimulus (CS). After a number of pairings of the meat powder (UCS) and the bell (CS), the bell develops the ability to elicit salivation on its own, the conditional response (CR), whereas formerly it would only elicit orientation to the sound

independently of meat powder. It was as if the paring with meat powder conferred a new meaning on the sound of the bell. A diagram of this experimental paradigm is shown in Fig. 6.1.

In the Pavlovian conditioning experiment, the UCS, which was meat powder that was blown into the dog's mouth, elicits salivation. Salivation elicited by the meat powder does not depend on any other event, and is therefore referred to as the unconditional response (UCR). The sound of the bell would normally elicit an orientation response from the dogs. After a number of pairings between the bell and meat powder, however, the bell developed the ability to elicit salivation, which in this case is referred to as the conditional response (CR) because its ability to elicit salivation depends on the pairing with the meat powder. The ringing bell had become a signal or *sign* of the immanence of meat powder. The acquisition of new meaning by formerly neutral stimuli is demonstrated by the increased interest and the change in behavior that they provoke after pairing with a biologically relevant stimulus. Work with pigeons shows why this inference is justified on the basis of the pigeon's behavior.

Classical conditioning studies in pigeons show that they will orient toward, approach and peck a lighted opaque plastic disk on the wall of the experimental chamber that signals a brief interval of food or water availability when the light is on. Tellingly, the type of pecking response to this CS is one that is appropriate for drinking when water is used as the biologically relevant stimulus (UCS), while it is the type of pecking that is appropriate for eating when food is used (Jenkins HM and Moore BR, 1973). The elicitation of a consummatory response that is appropriate to

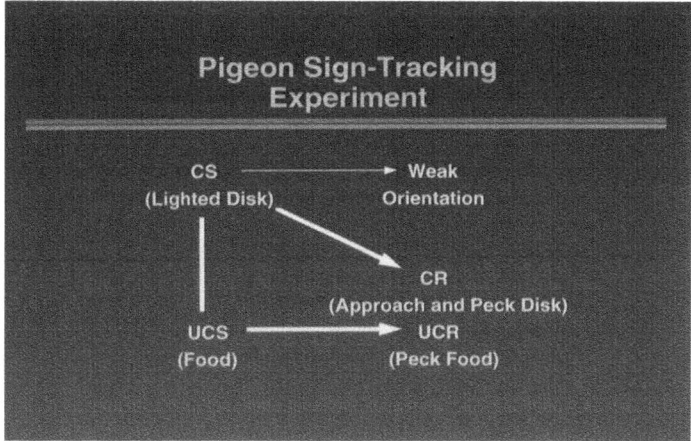

Fig. 6.2 The lighted disk acquires the ability to capture the pigeon's attention after it is paired with food, as demonstrated by the pigeon's orientation, approach and pecking

the type of UCS used clearly indicates that the pigeons relate to the lighted disk (CS) as a *sign* for that class of biological stimulus. The lighted disk acquires the ability to elicit the biologically appropriate response by virtue of the pairing with the biological stimulus. Clearly, the neutral stimulus becomes a *sign* for the biologically important one, and elicits intense interest after the pairings. An informational value is attached to a neutral stimulus (CS) after the pairings, which is different from the value assigned to it before. This phenomenon of orienting toward and approaching a stimulus that acquires significance by virtue of such pairing is known as *sign-tracking*. The acquired interest, approach and the appropriate type of pecking at the CS that pigeons display in sign-tracking provides strong behavioral evidence of the significance that organisms attribute to neutral stimuli that are correlated with biologically relevant ones. The diagram for the relationships between stimuli and responses is the same as that for Pavlov's dogs and is shown in Fig. 6.2.

Studies of *sign-tracking* such as this provide behavioral proof that the higher nervous systems of mammals and birds assign *meaning* to stimuli based on the *information* that is contained in stimulus relationships in the environment. The detection and assignment of meaning to stimuli and the relationships among them is taken to its highest level in the primates, especially in the most advanced member of that group, modern humans or *Homo sapiens sapiens*.

Evolution of the Modern Human Mind

Evolutionary Psychology is developing an understanding of the evolution of human cognition based upon the findings of physical anthropology, archeology, molecular biology, neuroscience and psychology. The species name for fully modern humans,

Homo sapiens sapiens, is evocative of the reflective quality that is essential to self-awareness if it is understood with some license as *Human who knows he knows.*[7] As famously expressed by Descartes, when he said I think therefore I am, *Homo sapiens sapiens* knows with epistemological certainty that the human mind is a *knowing* mind. We willingly attribute the same cognizance to others in the hypothetical construct known as *Theory of Mind*, in which we have confidence in making inferences about the mental and emotional states of others not only because we believe they think and emote in ways similar to ourselves, but also because we are capable of understanding that other humans can think and feel differently from the way we do. Self-reflective *knowing*, and the associated Theory of Mind, appear to be essential cognitive capabilities of our species. The origin of this *meta-awareness* must be understood if we are to explain the nature of the modern human mind and its capacity to contemplate the meaning of reality.

It is widely held among physical anthropologists that the evolution of modern humans began around five million years ago when the hominin antecedents of our species diverged from the last common ancestor shared with our closest living relatives the Chimpanzee. Figure 6.3 illustrates the geographic location for this evolutionary history, the Great Rift Valley of East Africa.

Figure 6.4 shows the timeline for hominin evolution together with illustrated lateral (side) views of the skulls of the different species involved.

Brain endocasts obtained from the skulls of representative hominins in the line leading to fully modern humans reveals a dramatic increase in brain size during the course of hominin evolution. This enlargement had a profound impact on the shape of the skull, particularly the slope of the forehead. The expansion of the brain did not affect all regions uniformly, however. More ancient parts of the brain, such as the Medulla and Hypothalamus that were responsible for regulation of basic body functions like breathing, heart rate and energy balance, were hardly affected at all, while the Neocortex which was responsible for higher aspects of information processing necessary for the growing faculty of abstract thought were affected the most. Figure 6.5 provides a schematic illustration of major brain regions.

To understand the significance of the evolutionary changes that led to the emergence of the modern human mind, we need to briefly consider the hierarchical organization of cortical information processing. Basic sensory information processing takes place in primary regions. Primary visual information processing occurs within the most posterior region of the occipital lobe, primary auditory processing takes place on the dorsal (top) region of the posterior part of the temporal lobe, and primary somesthetic[8] processing occurs in the most forward (anterior) region of the parietal lobe shown as somatosensory cortex in Fig 6.5a. Taste and

---

[7] I would like to thank a student in one of my classes at The University of Pennsylvania, Razeen Jivani, for calling my attention to this insightful interpretation of the scientific name for modern humans, *H. sapiens sapiens*.

[8] Somesthesis refers to the sensations of touch, temperature, pain, and proprioception (the sense of body position that derives from receptors in muscles, joints and tendons).

Fig. 6.3 The rift is a region in the African Continental Plate that is slowly splitting into two tectonic plates. This process is known as rifting and is caused by an upwelling of heat from the mantle into the overlying crust. As a result, the crust is separating and moving apart at a rate of 6–7 mm annually. Complete rupture of the African Plate is estimated to occur within 10 million years at which time a new ocean basin will form. (This public domain image is from: https://upload.wikimedia.org/wikipedia/commons/thumb/1/12/Gregory_Rift_Topographical.svg/2000px-Gregory_Rift_Topographical.svg.png and is used under terms at: https://commons.wikimedia.org/wiki/File:Gregory_Rift_Topographical.svg)

smell also have designated areas for basic sensory information processing. The gist of basic sensory information processing in these regions is that features of the holistic perceptual world, which is referred to as the perceptual Gestalt,[9] are broken down into the distinct elements of which the whole consists. Certain neurons in primary sensory cortices are activated optimally by these specific features. In this

---

[9]*Gestalt* is a German word that may be translated as form or shape. It conveys the idea that the objects of reality should be viewed holistically not as collections of parts. *Gestalt psychology* attempts to understand the ability to acquire and maintain meaningful perceptions of reality. Gestalt psychology was developed in the Berlin school of experimental psychology and has had a wide-ranging influence in diverse fields of psychology to impact the understanding of perception and clinical psychology in particular. Gestalt psychology continues to be an influence in cognitive neuroscience, especially in regard to what has been called the *binding problem*. The binding problem refers to the mechanism by which activity of neural networks that encode the diverse discrete component features of a percept are reintegrated into the Gestalt perception of a holistic object in the world.

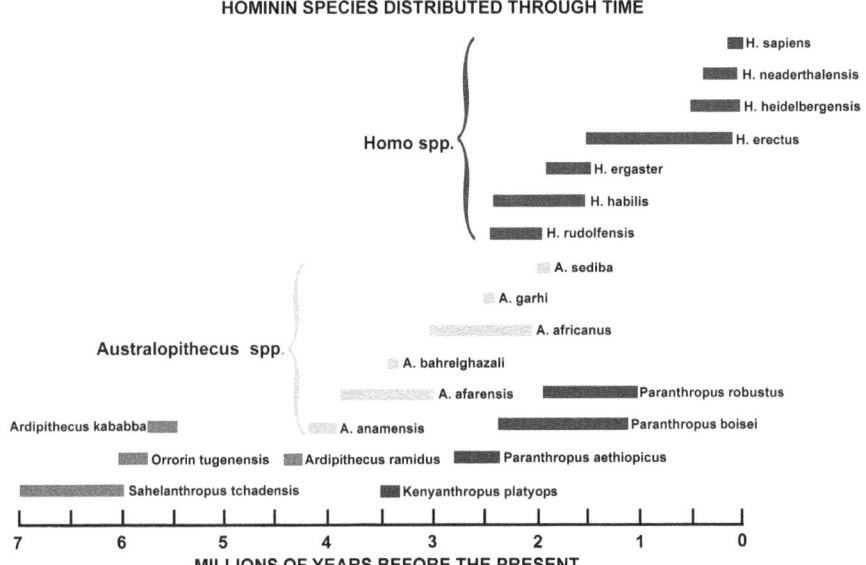

Fig. 6.4 The timeline of hominin evolution. The Australopithecines gave rise to *Homo habilis*, the first member of the *Homo* line that eventually led to *Homo sapiens*. The Australopithecines shown in green became extinct less two million years ago and were contemporaneous with several *Homo* species shown in blue. The sequence of species that gave rise to modern humans proceeds from *H. habilis* to *H. erectus* and then *H. antecessor* (not shown) which gave rise to *H. heidelbergensis*, the last common ancestor of archaic *H. sapiens* and *H. neanderthalensis*. *H. sapiens sapiens* (not shown) arose around 75,000 years ago. (Public domain image was downloaded from: https://upload.wikimedia.org/wikipedia/commons/b/ba/Hominin_evolution.jpg under Wikimedia Commons terms specified at: https://commons.wikimedia.org/wiki/File:Hominin_evolution.jpg)

manner networks of neurons are activated in which each class of neurons responds optimally to particular elementary features of the perceptual Gestalt. For vision, there are specific neurons in primary visual cortex that respond to lines at particular orientations that form an edge, and yet others that respond to two lines that form an angle. The excitation of this network of neurons provides the feature-information from which neurons in higher-level networks reconstitute the physical reality into synchronous neuronal activity to create the subjective perceptual experience of the physical *Gestalt*. In other words, primary visual cortex must first "deconstruct" an object into the neural representation of its component features and then higher order levels of the visual system "reconstruct" or integrate the neural code of component features into a holistic perception of an object in the world. The auditory and other sensory cortical areas do the same thing for their particular sensory modalities. More generally, each lobe of the cortex has distinct regions for this kind of basic and higher level information processing. The regions devoted to the higher-level functions are referred to as *association areas*. We will see below that distinct features of neuronal morphology, called dendritic spines, are specialized adaptations that enable

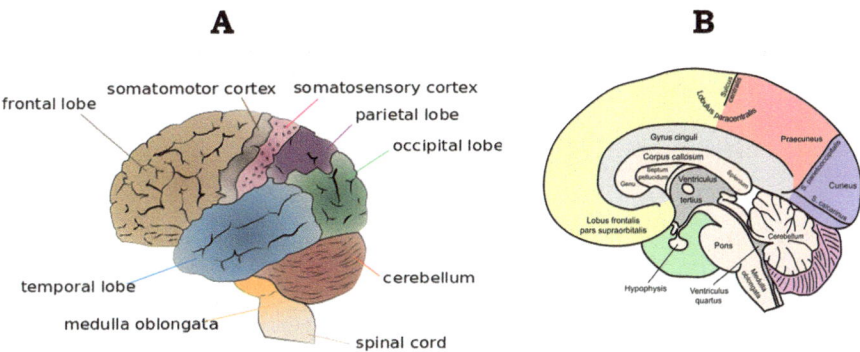

Fig. 6.5 (a) Schematic illustration of the lateral (side) view of the human brain which shows the left hemisphere. The neocortical surface is convoluted by sulci (grooves) and gyri (elevations). It is divided into distinct regions or lobes that specialize in different functions. Public domain image downloaded from: https://commons.wikimedia.org/wiki/File:Illu_cerebrum_lobes.jpg. (**b**) The medial (midline) surface of the right hemisphere is revealed along with sub-cortical brain structures. The area colored light pink is the brain stem, which consists of the Medulla, Pons and Midbrain (not labeled). The Cerebellum (pink and purple) sits above the Pons and Medulla. (Public domain image downloaded from: https://commons.wikimedia.org/wiki/File:Gehirn,_medial_-_beschriftet_lat. svg)

Table 6.1 Differential evolutionary expansion of selected brain regions of the human brain

Brain region	Factor in excess of expectation for non-human primates
Medulla	1.03
Diencephalon	1.6
Hippocampus	2.1
Cerebellum	2.8
Neocortex	3.2

Data from Donald (1991)

these different levels of information processing. With this background, we can return our attention to the evolution of the hominin brain.

As recounted by Merlin Donald, and shown in Table 6.1, hominin brain evolution had a differential impact on the expansion of different areas of the brain.

In general, the more ancient phylogenetic brain regions such as the medulla and midbrain were the least affected, while the more recently evolved areas such as the forebrain were the most affected. In the forebrain, the neocortex enlarged the most followed by the diencephalon, which contains the thalamus, hypothalamus and other structures. The evolutionary changes in the hominin brain clearly affected its more ancient regions that are concerned with basic functions such as heart rate, breathing and other vegetative functions, much less than the neocortex and related forebrain structures. The differential effect of evolution on brain structures was determined by how that structure compares in size to what would be predicted for non-human primates of the same weight. A value of 1.0 for the Medulla means that it is in line with the prediction, whereas the value of 3.2 for Neocortex means that humans have

Table 6.2 Approximate timelines for the succession of hominins in millions of years before present to show the chronology of anatomical and cultural change during hominin evolution

5.0 – Divergence of hominin and chimpanzee lines		1.5 – *Homo erectus*
4.0 – Australopithecines		**First major increase in brain size**
		More elaborate tools
		Use of fire and shelters
	Bipedal	Seasonal base camps
	Food sharing	First migration out of Africa
	Division of labor	**0.3 – Archaic *Homo sapiens***
	Nuclear family structure	
	Larger number of children	
	Longer weaning period	**Second major increase in brain**
2.0 – *Homo habiliis*		**size**
	As above plus stone tools	Modern form of vocal tract anatom
	Larger but variable brain size	**0.1 *Homo sapiens sapiens***
		Cave art, rituals, mythology

Information taken from Donald (1991)

3.2 times as much Neocortex as would be expected for a non-human primate of the same weight.

The neocortex experienced the greatest increase in size; and the prefrontal cortex, which is the part of the Neocortex that lies behind the forehead, expanded the most as implied by the change in the slope of the line drawn between eyebrow ridge and the frontal bones of the skull. The impact of the expansion of the neocortex and especially the prefrontal cortex on hominin cultural and technological change is summarized in Table 6.2.

A gradual increase in brain size began soon after the divergence of the hominins from the last common ancestor shared with the great apes. This is evident in the volumes of the cranial cavities of various australopithecines exemplified by the now famous *Lucy* and others. The trend in brain enlargement continued in the first member of the *Homo* line leading to modern humans, *Homo habilis*, which is the first species to demonstrate deliberate, but still crude, stone tool-making. The first and second major expansions of brain size occurred in *Homo erectus* and archaic *Homo sapiens*, respectively, and each of these expansions was accompanied by a corresponding technological advance indicated by the refinement of stone tools, as well as cultural adaptations for living in large groups, such as the use of seasonal base camps, the development of language, depictive art and ritual practices in modern humans. These findings indicate that the advances in brain size and the corresponding enhancements of brain computational power were connected to dramatic increases in general, technological and social intelligences.

The idea that the hominin brain had domains comprised of neural networks for different types of intelligence was put forward by Steven Mithen. Mithen elaborated on the idea that in the early hominins and their antecedents a general intelligence domain, and various specialized domains related to social, natural history, communication and technological intelligences, existed as functionally segregated capabilities of the brain (Mithen 1996). That is, early hominins did not have an effective neurological mechanism to correlate information from each of the diverse intelligence domains. Mithen postulated that the capacity for doing this emerged fully in early modern humans, and proposed a sequence of stages in the evolution of the hominin brain that led to the unique type of *Cognitive Fluidity* that we enjoy as fully modern humans. In *Stage I,* minds were dominated by a *General Intelligence* suite of general purpose learning and decision-making rules. In *Stage II,* General Intelligence was supplemented by multiple specialized intelligences that were each devoted to a specific domain of behavior and operated independently of each other. These supplementary intelligence domains are concerned with *Social, Technical, Linguistic and Natural History, Intelligences*. Finally, in *Stage III* the hominin mind advanced by the appearance of mechanisms for information exchange among formerly non-interacting intelligence neural networks. What neurological adaptation (s) provided the mechanism for cognitive fluidity that is enabled by the rapid exchange of information among the various neural networks responsible for the different types of intelligence? The runaway expansion of the cortex, especially the prefrontal cortex, provides a likely candidate.

As already mentioned, the prefrontal cortex expanded more during human evolution, in relative terms, than any other cortical region especially during the later stages of hominin evolution. The prefrontal cortex also has extensive connections with other cortical areas, as well as with more primitive sub-cortical brain regions. This raises the possibility that the extensive connections of the prefrontal cortex, together with its increased size and computational power, may have provided the mechanism needed for information from functionally segregated neural intelligence networks to be accessed and synthesized into novel modalities of thought. This would have produced new and dramatic cognitive capabilities that persist in contemporary humans as well. In addition to an increase in size, which implies an increase in neuron number, can we find evidence for a change in the internal structure or cytoarchitecture[10] of the prefrontal cortex? Area 10 of the prefrontal cortex is of special interest in this regard. Figure 6.6 shows the location of Area 10.

The information we are seeking is shown in Tables 6.3 and 6.4 below. Table 6.3 compares the whole brain and Area 10 volumes in humans and hominoids.

The analysis clearly showed that whole brain volume is much greater in humans as expected. The data also revealed that human Area 10 is much larger than that of the other hominoids. We expect on the basis of this result that Human Area 10 also would have more neurons than the other hominoids. The data on neuron number shown in Table 6.4 shows this to be the case. Of special significance, however, is the

---

[10]Cytoarchitecture refers to the size, shape, distribution and number of neurons in a particular brain region, as well as details about how those neurons are connected to each other and other neurons elsewhere in the brain. It is what is seen and quantified when looking through the microscope at a section of brain tissue.

Fig. 6.6 Lateral view of human brain contained within the skull. The dark red area corresponds to area 10 of the Prefrontal Cortex. (Public domain image at: https://upload.wikimedia. org/wikipedia/commons/7/ 7c/Brodmann_area_10_ lateral.jpg)

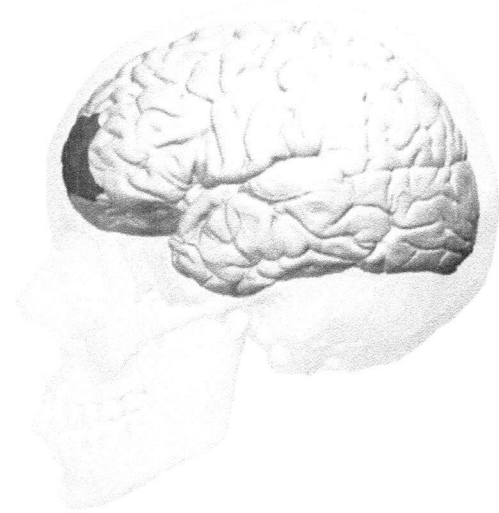

Table 6.3 Total right hemisphere brain volume, and right hemisphere Area 10 volume, for each hominoid species shown in mm^3

Species	Brain	Area 10
Human	**1,158,300**	**14,217.7**
Chimpanzee	393,000	2239.2
Bonobo	378,400	2804.9
Gorilla	362,900	1942.5
Orangutan	356,200	1611.1
Gibbon	88,800	203.5

Estimates were obtained from one individual in each species. The data are taken from Semendeferi et al. (2001)

Table 6.4 Estimates of total neuron number and neuron density in Area 10 in one hemisphere of humans and living Great Apes reveal a much higher number of neurons that are more widely spaced in humans

Species	Neuron density per mm^3	Total numbers
Human	**34,014**	**254,400,000**
Chimpanzee	60,468	64,500,000
Bonobo	55,690	63,500,000
Gorilla	47,300	45,900,000
Orangutan	78,182	63,000,000
Gibbon	86,190	8,000,000

Semendeferi et al. (2001)

fact that while humans have many more neurons in Area 10 the density of those neurons is much lower in comparison to other hominoids.

How might a lower density of a larger number of neurons in a larger volume represent an adaptation in Area 10 that supports the Cognitive Fluidity proposed by Steven Mithen? Figure 6.7 points the way toward our answer.

A   B

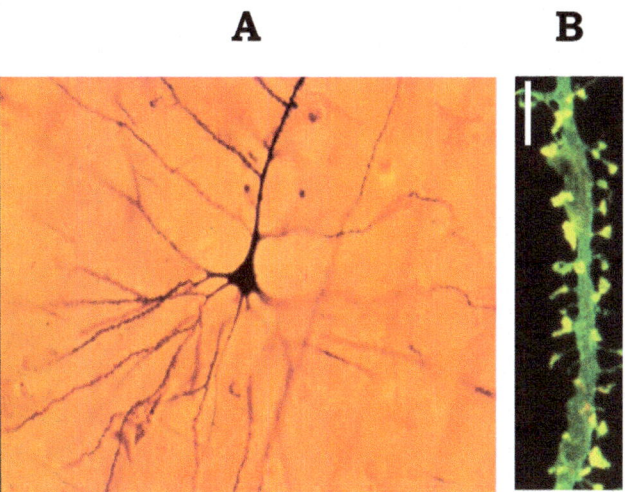

Fig. 6.7 (**a**) Golgi Stained Cortical Pyramidal Neuron. Note the pyramidal shaped cell body with the long thick apical dendrite rising to the top of the frame. Dendritic spines can be seen at this magnification as very small protrusions emanating from the apical dendrite. Basal dendrites are seen emanating from the bottom of the cell body and extending toward the bottom of the frame. Photo by Bob Jacobs, Laboratory of Quantitative Neuromorphology Department of Psychology Colorado College. This public domain image was downloaded from https://upload.wikimedia.org/wikipedia/commons/6/6d/GolgiStainedPyramidalCell.jpg and is used here under a license found at: https://en.wikipedia.org/wiki/GNU_Free_Documentation_License. (**b**) Immuno-Fluorescence image of a horizontally oriented dendrite (shown in green) with multiple protruding dendritic spines (shown in yellow). High spine density and many spines with long necks are typical characteristics of human pyramidal neurons. The vertical bar = 5 μm. The image was downloaded from https://upload.wikimedia.org/wikipedia/commons/2/26/Cytoskeletal_organization_of_dendritic_spines_%28ru%29.jpg and is reproduced under terms of the Creative Commons Share Alike license at: https://en.wikipedia.org/wiki/Creative_Commons. (The image originally appeared in Hotulainen and Hoogenraad 2010)

As shown in Fig. 6.7a, pyramidal neurons have extensions of the cell body called dendrites. The neuron of interest is shown in the figure to have a long dendrite that arises from the apex of the triangular (pyramidal) cell body. The apical dendrite has branches and, together with the basal dendrites that emanate from the other sides of the cell body, comprises the dendritic extensions collectively referred to as the *dendritic tree* or *arbor*. The ultrastructural feature of interest is shown Fig. 6.7b. This fluorescent image of a dendrite shows small extensions called dendritic spines. These are target regions onto which other neurons make synaptic contact to activate or inhibit the target neuron. It has a presynaptic component that is the termination of an axonal extension that arises from another neuron, and a postsynaptic component which is the membrane of the dendritic spine itself. The small gap is the synaptic cleft. Figure 6.8 illustrates these features of a synaptic contact on a dendritic spine.

With this additional information, we can return to consider the significance of the unique cytoarchitectural features of human prefrontal cortex Area 10. The larger number of *less densely packed* neurons in human Area 10 compared to other species

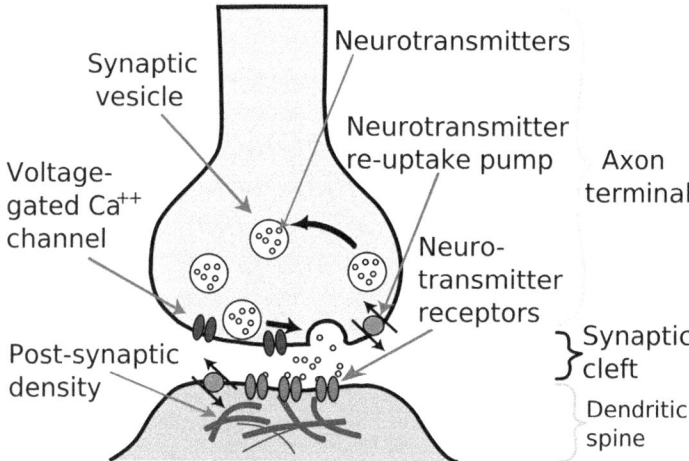

Fig. 6.8 Schematic of an axon terminal making a synaptic contact on the head region of a dendritic spine. The axon terminal shows small vesicles that contain neurotransmitter substance that will be released when an action potential reaches the axon terminal. After release, neurotransmitter diffuses across the synaptic cleft and binds with receptors on the spine head. This produces a change in the electrical potential (voltage) across the spine membrane. Excess neurotransmitter is taken up from the synaptic cleft by transporter molecules in the membrane of the axon terminal and is then repackaged in vesicles to be released again with the arrival of subsequent action potentials. Synapses on pyramidal neurons are excitatory, but other neurons can have both excitatory and inhibitory synapses. Depending on whether the neurotransmitter is excitatory or inhibitory the voltage across the membrane will either decrease or increase, respectively. This makes it easier or more difficult for the postsynaptic neuron to be activated by other neurons. (Image at: https:// upload.wikimedia.org/wikipedia/commons/thumb/e/e0/Synapse_Illustration2_tweaked.svg/ 1280px-Synapse_Illustration2_tweaked.svg.png. The image originally appears in (Julien 2005) and is used here under a license found at: https://creativecommons.org/licenses/by-sa/3.0/deed.en)

allows for more and longer dendrites, as well as more dendritic spines on each dendrite. The presynaptic elements for the synapses that form on dendritic spines in Area 10 are axon terminals of neurons that have their cell bodies in cortical association areas in the other lobes of the cortex, as well as in other regions of the prefrontal cortex. These association areas are themselves specialized for higher levels of information integration as explained above in relation to the brain anatomy illustrated in Fig. 6.5. The prefrontal cortex is exclusively devoted to association functions. Area 10 of the prefrontal cortex, in particular, appears to be a highly-specialized association area that is exclusively dedicated to the integration of information from other "lower-level" association areas, including other regions within the prefrontal cortex. The evolutionary changes in Area 10 under consideration here are thought to have been enabled by duplication of certain genes in hominins during the time that the *Homo* line was being established with the divergence of *H. habilis from the Australopithecines*. One such gene, designated SRGAP2, duplicated to create a variant called SRGAP2B about 3.4 million years ago, long after the hominins diverged from the last common ancestor shared with the

Chimpanzee. About one million years later SRGP2B duplicated to form SRGAP2C, a shortened version of the original (Denis et al. 2012). SRGAP2C slows pyramidal neuron differentiation to allow more dendritic spines to form (Charrier et al. 2012). The increased number of dendritic spines would provide more potential targets for incoming axons seeking an opportunity to form a synaptic union during the neurodevelopmental process called synaptogenesis. This change would allow pyramidal neurons in human prefrontal cortex, especially those in Area 10 that seem to have been most affected, to specialize in receiving and integrating more synaptic inputs from a greater number of other neurons. This possibility is certainly consistent with the larger more broadly ramified dendritic tree, longer apical dendrites and more synaptic spines per unit dendrite length in human Area 10.

Other gene duplications and mutations likely are involved in the emergence of the modern human mind, however, because the last known duplication of SRGAP2 occurred about 2.4 million years ago. This set the course, but the fully modern condition of the human brain was realized approximately 100,000 years ago. In the time between the last duplication of SRGAP2 and the emergence of the modern human mind it is probable that changes in neurogenesis continued to produce progressive increases in the number of cortical neurons generated during brain development. The human power of abstraction is evident in the artifacts and cultural innovations of *Homo erectus*, but it seems to have remained relatively static for over a million years until some additional change(s) gave rise to the emergence of archaic *Homo sapiens* and then finally fully modern humans who we have been referring to as *Homo sapiens sapiens*. What were they like? To answer that question, we only need to look into a mirror. These people erupted from Africa in what eventually became a global migration. A population consisting of thousands has grown to well over seven billion who now permanently inhabit most of the Earth's surface.

The Advent, Achievements and Perils of Modern Humanity

Humanity has many impressive accomplishments to its credit, especially since the advent of modern science and technology, but we stand in grave peril of self-inflicted harm on a global scale that results primarily from a failure to understand our own emotional and cognitive processes. Our ancestors came out of Africa with vastly increased powers of abstraction that enabled the emergence of superior weapons technology and hunting strategies, depictive art and ritual practices that suggest a strong desire, even compulsion to understand and explain the world. We carry their genes, so we must see and understand the manifestation of those Pleistocene ancestors in ourselves.

The enhanced cognitive capabilities of archaic and modern humans introduced the capacity to magnify fear as well as master the environment. What humans do not understand, therefore, has great potential to generate fear. One only has to consider the powerful influence of thoughts on breathing and heart rate in panic disorder to appreciate how a lack of understanding can run amuck to generate all of the autonomic symptoms of intense fear. Furthermore, sapience led to the realization of what is for many the ultimate source of fear, the awareness of one's own mortality; and so a certain existential dread is part and parcel of sapience. Christian theologian, Paul Tillich, wrote persuasively in his book, "The Courage to Be", about the origin of existential anxiety in humanity's fear of non-being in death (Tillich 1952):

> The fear of death determines the element of anxiety in every fear. Anxiety, if not modified by the fear of an object, anxiety in its nakedness, is always the anxiety of ultimate nonbeing. . . . It is the anxiety of not being able to preserve one's own being which underlies every fear and is the frightening element in it. In the moment, therefore, in which "naked anxiety" lays hold of the mind the previous objects of fear cease to be definite objects. They appear as what they always were in part, symptoms of man's basic anxiety. . . . The basic anxiety, the anxiety of a finite being about the threat of nonbeing, cannot be eliminated. It belongs to existence itself.

In their insightful book, "Why God Won't Go Away", neuro-radiologist Andrew Newberg and his colleagues (Newberg et al. 2001) consider existential dread, which is evoked by the sense of mortality, in the minds of *Neanderthals* and our modern human forebears, archaic *Homo sapiens*.

> Somewhere in the mists of human pre-history, our slope-browed Stone-Age cousins, now known as Neanderthals, apparently became the earth's first living creatures to bury their dead with ceremonies. We can only imagine what dark thoughts possessed those gruff and shaggy nomads as they lay their clan mates to rest . . . .the graves had been carefully provisioned with weapons, clothing and other essential supplies. Perhaps these were gifts. . . .More likely, it seems our Neanderthal progenitors were outfitting their dead with gear to help them meet whatever mysterious adventures lay ahead.

> This poignant optimistic gesture – histories first-known glimmering of metaphysical hope – tells us two important things about our Neanderthal ancestors: first, that they possessed sufficient brain power to comprehend the inescapable finality of physical death; and second, that they had already found a way to defeat or cope with it at least conceptually. . . . .

> The graves and shrines of the Neanderthals are the earliest known evidence of proto-religious behavior. The fact that they occur coincidentally with the earliest evidence of human culture. . . . . . .suggests something important: As soon as hominids began to behave like human beings, they began to wonder and worry about the deepest mysteries of existence – and found resolutions for those mysteries in the stories we call myths.

Several pages later, the authors continue,

> The cognitive imperative drives the higher functions of the mind to analyze the perceptions processed by the brain and transform them into a world full of meaning and purpose. By doing so, it has given human beings an unsurpassed genius for adaptation and survival. But these cognitive abilities have a down side as well. In its tireless quest to identify and resolve any threat that can potentially harm us, the mind has discovered the one alarming apprehension that can't be resolved by any natural means – the sobering understanding that everyone dies. . . . . . .This grim discovery must have entered the world soon after self-awareness began to glow in some prehistoric human mind. . . . . .But death was not the only existential worry that early humans had to face. By comprehending their own mortality, they had stumbled onto a new dimension of metaphysical worries, and their questioning minds must have presented them with difficult and unanswerable questions at every turn: Why were we born only to eventually die? What happens to us when we die? What is our place in the universe? Why is there suffering? What sustains and animates the universe? How was the universe made? How long will the universe last?

There is evidence of interbreeding between *Neanderthals* and *Homo sapiens sapiens* in the Middle East soon after modern humans left Africa, based on the finding of some elements of the *Neanderthal* genome in modern humans of European and Asian descent (Green et al. 2010). Both species lived contemporaneously in Europe for thousands of years after modern humans arrived on the European continent. Although we carry some of their genes, modern humans did not descend from Neanderthals. The likely common ancestor of *H. Neanderthalensis and* archaic *Homo sapiens* is *Homo heidelbergensis*, a late variant of *Homo erectus* via *H. antecessor*, who also may have buried their dead at least in an effort to prevent scavenging by carnivores. The anthropological evidence of mortuary ritual practiced by Neanderthals is similar to what Newberg and his co-authors described. A 12-year re-analysis of a Neanderthal burial site at La Chapelle-aux-Saint supports the idea of intentional burial by Neanderthals (Rendu et al. 2014.) This site was first discovered in 1908, but the question concerning whether *Neanderthals* practiced intentional burial of their deceased remained controversial until the extensive re-analysis of the site. A meta-analysis of known Neanderthal burial sites supports the thesis of intentional Neanderthal burial with ritual features related to corpse processing (de-fleshing and disarticulation), non-nutritional cannibalism, and curation of body parts (Schwarz 2013). Schwarz draws interesting conclusions concerning the mental state of the mourners, their understanding of the meaning of death based on an empathetic effort to forestall death in an injured arthritic elder of the group:

> As social creatures Neanderthals would have created close relationships within hunter-gatherer groups, which would result in a traumatic experience at the sudden loss of one of these individuals. Neanderthals have demonstrated this understanding in their attempts to delay the onset of death in elderly group members, such as the individual known as The Old Man of Shanidar (Shanidar 1, Shanidar Cave, Iraq).21 This elderly and wounded Neanderthal man would have required significant care from the group to survive on a daily basis with such significant injuries for such a lengthy period of time. This demonstrated that Neanderthals had recognised that death would result from a lack of support, so they empathised and concluded that this would be a negative result, and that the group may suffer as a result (either through a negative psychological, social, or physical impact).... In order to have empathy for another, we have to be able to place ourselves in the mind of the other individual.

Caring for infirm group members coupled with the practice of mortuary ritual in the care for corporeal remains is suggestive of a capacity for empathy, which in turn requires a developed sense of self and theory of mind. The inclusion of personal objects as part of burial ritual seems to have been a more recent innovation practiced primarily by modern humans. Relatively few findings of provisioned grave goods in Neanderthal burials makes it more difficult to infer belief in an afterlife among archaic *Homo sapiens* such as *Neanderthals*. This type of provisioning of the dead shows awareness that death represents a change or transition in which the self is separated from the body, and a belief or wish that death does not annihilate the self. Burial ritual is therefore an expression of *metaphysical hope* according to Andrew Newberg and his colleagues (Ibid). In any event the appearance of metaphysical-existential awareness, as evidenced by mortuary ritual in modern humans and perhaps the *Neanderthals*, is intimately associated with the evolution of human self-awareness. In his book on human consciousness, "Wider than the Sky" (Edelman 2004), Gerald Edelman speculates that a form of primitive self-awareness, which he calls primary consciousness, first emerges in birds and mammals when they diverged from reptiles. The consciousness of *Homo sapiens sapiens* evolved to a substantially "higher" level based on antecedents that first appeared in earlier hominins but especially *H. heidelbergensis*, the likely common ancestor of modern humans and *Neanderthals*. This evolution reaches its most advanced state in modern humans who attempted to address their existential fear in the context of proto-religious ritual and associated mythological story-telling.

Early humans first attempted to reduce fear of the unknown, including existential fear related to the awareness of mortality, through the invention of myths. These stories purported to explain the origin of the world and the place of humans in it, as well as their relationship to supernatural beings, spirits and gods, who controlled the forces of nature. The use of mythology for this purpose is a manifestation of what Newberg et al. (Ibid) refer to as the *cognitive imperative*, which appears to be an essential feature of modern human sapience. Humans feel compelled in varying degrees to understand and explain phenomena that they observe, even if the "explanations" are not rational. The proto-religious hypothesis formation involved in the myth-making of early modern humans very likely was driven by the imperative to understand and manage existential fear, and more generally fear of the unknown. Likewise, the stories that young children invent to explain imaginary events, how something happens, or even their own behavior, may be an example of the same process in which magical thinking is presented as having explanatory power. The magical thinking of contemporary children is used to create imagined constructs of reality that represent the "story" of how they experience their world. The drive to explain reality in terms of hypothetical constructs persists not only in contemporary children, but also adults in general and scientists in particular.[11] This concept was captured by psychologist, George Kelly, in his personal construct theory of personality (Kelly 1955). Kelly proposed that everyone is a scientist insofar as all individuals construct explanations for their perceptions, experiences and social interactions, as well as a personal understanding of reality.

The problem with relying on hypothetical constructs of reality, whether they are in the form of myth, religious dogma, political ideology or scientific hypothesis as a way to diminish existential fear, and fear of the unknown, is that it motivates people to overstate and overvalue the ideas, hypotheses or beliefs that they promulgate. There is a great predilection to state one's ideas or beliefs in absolute terms as representing the certain truth[12] about some aspect of reality, and a corresponding fear that if a belief is proven wrong the protection that it provides against the unknown

---

[11] As the brain matures, especially the prefrontal cortex, these hypothetical constructs become more reality based, but this is not always true in regard to inferences made concerning the thoughts and emotions of others in Theory of Mind constructs. Varying degrees of paranoia about the intentions of others are essentially errors in Theory of Mind constructs.

[12] In the next chapter, we will see that the apprehension of certain truth can only be achieved in the limited case of deductive logical proofs argued from true premises.

will be lost as well. Hence, we humans tend to defend our personal beliefs far beyond the point where they are rationally defensible. This is one of the key factors that lead to the polarization of argument, especially concerning the big questions related to the meaning of life, existential fear, and the question of God's existence. For this reason, humanity has shown from the beginning that it is willing to kill to defend ideas; and this is an unfortunate and unique invention of our species that derives from the emergence of sapience.[13]

As we have seen in previous chapters, *H. sapiens sapiens* has made tremendous progress in understanding the nature of the universe and our place in it using the methods of science. While we have developed a sophisticated and profound understanding of nature and its laws, our current state of knowledge does not yet address or remove the existential fear that has been the bane of humankind since the dawn of self-awareness. Moreover, science may never provide an answer to the ultimate questions concerning the origin of the universe and the existence of God. Can sapient mind formed by natural processes in the universe ever answer these questions? Can logic and philosophy fill this void? These questions are tackled in the next chapter.

References

Caporale LH (2003) Darwin in the genome: molecular strategies in biological evolution. McGraw-Hill, New York, p 13

Charrier C et al (2012) Inhibition of SRGAP2 function by its human-specific paralogs induces neoteny during spine maturation. Cell 149:923–935

Denis MY et al (2012) Evolution of human-specific neural SRGAP2 genes by incomplete segmental duplication. Cell 149:912–922

Donald M (1991) Origins of the modern mind. Harvard University Press, Cambridge, MA, p 108

Edelman G (2004) Wider than the sky. Yale University Press, New Haven

Gould SJ, Eldredge N (1977) Punctuated equilibria: the tempo and mode of evolution reconsidered. Paleobiology 3(2):115–151 p145

Green RE et al (2010) Divergence of Neandertal and human genomes. Science 328:710–722

Hotulainen P, Hoogenraad CC (2010) Actin in dendritic spines: connecting dynamics to function. J Cell Biol 189(4):619

---

[13]It is perhaps no accident that the biblical allegory of Cain and Abel follows quickly on the heels of their parents eating the forbidden fruit of the tree of knowledge. In this story, Cain's motive for killing his brother centered on what was essentially a theological dispute. Cain's belief that Abel's sacrifice was judged to be more-worthy and pleasing to God led to envy, anger and fratricide, because Cain feared the loss of God's favor and the protection it would provide. The echo of this primal crime resonates throughout history in many examples of ideologically driven mass persecution and murder inspired and implemented by narcissistic and psychopathic leaders, such as Adolf Hitler, Joseph Stalin, Idi Amin, and others who exploit differences in religious or political beliefs to further their own unconscious malignant agendas. The wisdom of humanity lies in the use of the enhanced cognitive skills that confer sapience on our species to understand reality and reduce fear of the unknown. The "original sin" of humanity is rooted in the overvaluation of the ideas and beliefs that sapience generates and the corresponding fear that competing ideas pose a mortal threat.

Jenkins HM, Moore BR (1973) The form of the auto-shaped response with food or water as Reinforers. J Exper Analysis of Behav 20:163–181

Julien RM (2005) A primer of drug action: a comprehensive guide to the actions, uses, and side effects of psychoactive drugs worth publishers, New York, pp 60–88

Kandel ER (2000) The molecular biology of memory storage: a dialogue between genes and synapses. Nobel Prize Lecture

Kelly GA (1955) The psychology of personal constructs. Norton, New York

Koonin EV (2007) The biological big bang model for the major transitions in evolution. Biol Direct 2:21

Kyriakis JM, Avruch J (2001) Mammalian mitogen-activated protein kinase signal transduction pathways activated by stress and inflammation. Physiol Rev 81(2):807

Maier NRF, Schneirla TC (1935) Principles of animal psychology. Dover Publications, New York, pp 7–35

Mithen S (1996) Prehistory of the mind: the cognitive origins of art and science. Thames and Hudson, London

Newberg A, D'Aquill E, Rause V (2001) Why god won't go away. Ballantine Books, New York, pp 54–55

Parkinson JS (1993) Signal transduction schemes of bacteria. Cell 73:657–871

Penrose R (2011) Cycles of time. Alfred A. Knopf, New York, pp 77–79

Rendu W et al (2014) Evidence supporting an intentional Neandertal burial at La Chapelle-aux-Saints. PNAS 111(1):81–86

Schrödinger E (1944) What is life. Cambridge University Press, Cambridge, pp 81–93

Schwarz SM (Winter 2013–2014) The mourning dawn: neanderthal funerary practices and complex responses to death. HARTS Minds J Humanit Art 1(3):1–14

Semendeferi K, Armstrong E, Schleicher A, Zilles K, Van Hoesen GW (2001) Prefrontal cortex in humans and apes: a comparative study of area 10. Am J Phys Anth 114:224–241

Tillich P (1952) The courage to be. Yale University Press, New Haven, pp 38–39

Chapter 7
Mind Knowing Truth

Richard J. Di Rocco

> *I don't know what I may seem to the world, but, as to myself, I seem to have been only like a boy playing on the seashore, and diverting myself in now and then finding a smoother pebble or a prettier shell than ordinary, whilst the great ocean of truth lay all undiscovered before me.*
>
> Sir Isaac Newton in a letter written shortly before his death
>
> *"What is truth?"* (Question of Pontius Pilate at the trial of Jesus.)
>
> Jn. 18:38
>
> *If you don't care where you are going, any road will take you there* (This is a common paraphrase of a longer passage from "Alice in Wonderland": "Would you tell me, please, which way I ought to go from here? That depends a good deal on where you want to get to, said the Cat. I don't much care where—said Alice. Then it doesn't matter which way you go, said the Cat. —so long as I get SOMEWHERE, Alice added as an explanation. Oh, you're sure to do that, said the Cat, if you only walk long enough.")

Abstract How is truth discovered and proved? One approach to answering this question begins by focusing the question more narrowly to ask whether new mathematics is invented or discovered. Opinions differ on the answer to this question, but everyone agrees that "new mathematics" must be expressed in the form of true mathematical statements. Are statements of mathematical truth discovered or invented? When the question is posed this way, it is clear that mathematical

R. J. Di Rocco (✉)
Psychology Department, University of Pennsylvania, Philadelphia, PA, USA

Psychology Department, St. Joseph's University, Philadelphia, PA, USA
e-mail: richdi@upenn.edu

truth may be discovered, but never invented. This follows from the fact that true statements can be proved using deductive reasoning that leads from a statement already known to be true to the hypothesized mathematical truth which is to be proved. The key point about a sequence of deductive logical statements is that the conclusion is *necessarily* true if the premises of the argument are true. Mathematical truth, indeed all truth, must exist *before* it is discovered. It exists eternally and cannot be invented. In what does necessary eternal truth have its existence, in what does it subsist? The answer to this question has profound implications for the search for ultimate meaning. The misapprehension that new mathematics is invented derives from the impression of a creative event when new ideas arise in the mind of the mathematician by spontaneous insight. Spontaneous insight will be examined as the end-product of inductive reasoning. What philosophers, psychologists and neurobiologists have said about inductive reasoning will be explored at length in this chapter in an effort to answer the questions, "what is truth" and "how is it discovered and proved".

Keywords Deductive reasoning · Inductive reasoning · Spontaneous insight · Epistemology · The nature of inquiry

Is New Mathematics Invented or Discovered?

In the pursuit of truth, we care where we are going. The problem is to find the road that takes us there when there are an unlimited number of directions upon which to embark. When confronting a problem, what path of inquiry will lead to the solution? Which way does one head in pursuit of the truth? The field of mathematical inquiry provides a useful context within which to examine this issue. Consider this question: *do mathematicians invent or discover new mathematics?* This is an important question because, as we shall see, the answer has potentially far-reaching implications. Mathematicians and philosophers have come down on both sides of the question concerning whether new mathematics is discovered or invented. This divergence, however, results from a failure to take account of the interplay between inductive and deductive reasoning in mathematical inquiry, as explained below.

New mathematics may involve the introduction of a new mathematical object or concept that is no more than an extension of one that already exists. This was certainly true for the introduction of the number zero, which was an extension of the *natural*, or counting, numbers (1, 2, 3, 4, ... N) to represent the absence of quantity. The concept of the *natural numbers* as a measure of quantity and the notion of *no quantity* were always present to the human mind. Introduction of the number zero was a formalization of the concept *no quantity* that nevertheless had far reaching consequences for arithmetic. The introduction of the negative whole numbers extended the number concept beyond the counting numbers and zero to form the set of numbers called *Integers*. The integers further extended the range of arithmetic operations. Likewise, the *set* concept is no more than the formalization of the idea of a collection of related things defined by some criterion of inclusion. For example, we have just used the set concept implicitly to refer to the collection of the counting numbers, zero and the negative whole numbers and give this set the name *Integers*. This usage is extremely intuitive and is readily understood, yet set theory had a major impact on developing firm foundations for mathematics in general, as well as calculus, topology, probability theory and abstract algebra in particular. Such extensions of existing mathematical objects, or concepts, usually open new areas of mathematical inquiry because they allow mathematicians to do things they could not do before. Nevertheless, such novelties are incremental and may lead only to incremental new mathematical results. On the other hand, there are wonderful moments in the life of a mathematician when she or he imagines what appears to be an entirely new type of mathematics that introduces new mathematical objects, or an entirely new way of manipulating known mathematical objects. Such insights are not merely incremental. They are paradigm-shifting and disruptive new ways of seeing and doing mathematics. The introduction of algebraic group theory by Evariste Galois and Niels Henrik Abel is one example that is recounted with fascinating biographical information, as well as mathematical exposition, by Mario Livio in his book, "The Equation that Couldn't Be Solved (Livio M 2005). The Abel Prize is named after Abel, who like Galois tragically died young.

Whether incremental or paradigm shifting, mathematical innovations are stated in new hypotheses that are imagined theorems, or propositions, of putative truth. At this stage, the propositions have not yet been proved. New mathematical theorems or logical propositions must be proven to be true before they can gain general acceptance in much the same way that the hypotheses of science require experimental validation. There is an undeniable element of creativity and spontaneous insight that plays a vital role in mathematics. Yet this provides what *only appears to be an inventive step* in the formulation of a new mathematical theorem, because the apprehension or formulation of the theorem occurs as a sudden, spontaneous and often unexpected leap of insight. For this reason, there is a sense in the mathematician's mind that something vital and new has been glimpsed, imagined, or even *invented*. It nevertheless remains a mere putative truth until it can be proven, and this typically involves elements of deductive reasoning that justify the steps of the proof. Every high school geometry student knows, for example, that it is possible to prove a theorem when the laws of deductive reasoning are applied to certain "given" statements or axioms that are known or accepted, *a priori*, to be true. What is proved in this case could be a theorem that the geometer has imagined might be true and interesting. It is usually possible to find a "path" from some known truth, referred to as a *given* truth, to the stated theorem *provided that the theorem is in fact true*. Such a path exists if the theorem is true. Deduced truth is *necessarily* true if the premise of its argument is true! It therefore *exists* and cannot be invented. Truth is therefore *discovered* when the laws of deductive reasoning blaze a trail from some known truth to prove the hypothesized truth or theorem. The notion of invention in mathematics derives from the role that *inductive reasoning* plays in the musings that allow the mathematician to imagine new theorems, or find a key step in the deductive path that leads from known truth to proven truth from among the many blind alleys that lead nowhere. So, the essence of the question posed above about new mathematics can be reformulated as follows, "do mathematicians invent or discover mathematical truth". When the question is posed in this manner two things are immediately obvious. If mathematical truth is invented, humans have the power to create truth; but if it is discovered, then truth necessarily exists before it dawns in the human mind.

Generalization of the foregoing makes it clear that a vast web of true statements exists concerning everything that can be known. These true statements can be thought of as comprising the *set of true statements* that contains not only statements of logic and mathematical truth, but also every other truth that can be known. We can separate the *set of true statements* into two subsets along the lines specified above. The first subset consists of all true statements of logic and mathematics, and the second subset consists of all other true statements about the tangible universe or multiverse. In regard to the first subset, which consists of all true statements of logic and mathematics, we have seen above that its true statements are necessarily true. They exist before they are discovered. This follows from the fact that each of these truths is connected to every other one by some sequence of deductive logical statements called a proof or syllogism. A syllogism begins with a statement that is known, a priori, to be true and proceeds according to the laws of logic to successive true statements that lead ultimately to the truth that was to be proved. The latter truth may have been imagined as an unproven mathematical or logical theorem, the truthful status of which was established by the syllogism. It is much like trying to cross a stream by stepping on rocks along the way. Each of the rocks is real (true), and ultimately by the correct sequence of steps one arrives at the other bank of the stream, which was to be reached. We know all this because true statements are never contradictory. We know a priori, therefore, that a definite path of correct deductive reasoning can take us from any one truth to any other. Unfortunately, it is often unclear how to find the most direct, or even any, path of deductive reasoning that will take us there. It may also be unclear which specific statements of deductive reasoning justify the steps along the way from one truth to another. Trial, error and intuition help the inquirer find the direction of the most efficient path, and often also help to find the sequence of deductive steps, much as they help to find the correct sequences of turns that constitute the way out of a maze. In view of the necessity of mathematical truth proved by the laws of deductive reasoning, it is reasonable to answer the question posed at the beginning of this chapter by concluding that new mathematics (mathematical truth) is discovered, not invented.

Can the same be said of the second subset of the set of all true statements, that is, the set that consists of all true statements about the tangible universe-multiverse? These truths, like mathematical and logical truths, are hypothesized and their veracity is also discovered, not by syllogistic argument but by the syllogistic aspect of experimental design. These truths are discovered by the methods of rational empirical science. Scientists construct hypotheses by inferring a general principle or hypothesis about some aspect of reality from several particular observations using inductive reasoning. They then construct experiments in a way that uses logical assumptions (deductive reasoning) about why different groups or conditions would be expected to lead to different experimental results. It is important to note that, while deductive reasoning is used to construct an experimental design that will lead to different outcomes for different experimental groups, an experiment must be conducted to decide the outcome. We see in rational empirical science the same interplay between inductive and deductive reasoning that we encountered in the mathematician's imagining of a new hypothesis (mathematical theorem) and its subsequent syllogistic proof. That is where the similarity in the methods of discovering logical-mathematical truth versus empirical truth ends, however. For, while mathematics is the language of nature, nature is far more complex than mathematics and logic in the sense that more variables need to be taken into account in assessing natural phenomena in contrast to mathematical theorems. As a result, hypotheses about the workings of the natural world are sometimes more difficult to prove and are more prone to error and incompleteness. For this reason, we find that empirical science typically reveals only limited apprehension of the truth about the natural world, while mathematics and logic lead to the discovery of definitive mathematical and logical truths providing that the premises of the syllogistic arguments by which they are proved are true.

Inductive Reasoning and the Spontaneous Nature of Insight

What is inductive reasoning and how does it operate? The process of scientific induction was elucidated clearly by Sir Francis Bacon in his seminal work on the scientific method, *Novum Organum*. Bacon was disinclined toward the syllogism as a method of discovering general truths about nature. According to this entry in the Stanford Encyclopedia of Philosophy (Jürgen K 2015), for Bacon the syllogism was inferior to the method of inquiry that he championed:

> Already in his early text Cogitata et Visa (1607) Bacon dealt with his scientific method, which became famous under the name of induction. He repudiates the syllogistic method and defines his alternative procedure as one "which by slow and faithful toil gathers information from things and brings it into understanding" (Reference Omitted). When later on he developed his method in detail, namely in his Novum Organum (1620), he still noted that

> > [of] induction the logicians seem hardly to have taken any serious thought, but they pass it by with a slight notice, and hasten to the formulae of disputation. I on the contrary reject demonstration by syllogism ... (reference Omitted).

To Bacon's point regarding the implied superiority of induction over syllogistic (deductive) argument, it should be noted that induction leads to conjecture about putative truth while the deduction leads to proof of truth! Nevertheless, understanding that empirical science requires the formation of hypotheses about a phenomenon that could then be tested experimentally, Bacon sought an inductive method to facilitate hypothesis formation. He was remarkably successful in this matter, and imitated by many successors. Bacon's scientific method consists of observation and systematic curation of data, which if done properly would lead to induction of hypotheses (axioms)[1] to explain those data. The hypotheses are then tested

---

[1]Bacon's use of the term *axioms* as the name for an inferred hypothesis reflects his understanding that inductive reasoning produces a proposed *general truth* about a phenomenon.

experimentally and either validated or refuted. Specifically, Bacon advocated a procedure for induction in which as many instances as possible of a phenomenon under consideration be listed along with counter-examples and then examples of variation in the phenomenon of interest. According to the Stanford Encyclopedia of Philosophy (Ibid):

> Induction implies ascending to axioms, as well as a descending to works, so that from axioms new particulars are gained and from these new axioms. The inductive method starts from sensible experience and moves via natural history (providing sense-data as guarantees) to lower axioms or propositions, which are derived from the tables of presentation or from the abstraction of notions. Bacon does not identify experience with everyday experience, but presupposes that method corrects and extends sense-data into facts, which go together with his setting up of tables (tables of presence and of absence and tables of comparison or of degrees, i.e., degrees of absence or presence). "Bacon's antipathy to simple enumeration as the universal method of science derived, first of all, from his preference for theories that deal with interior physical causes, which are not immediately observable" (Reference Omitted)

To summarize, while deductive reasoning argues from a given general truth to prove a derivative particular truth, inductive reasoning does the opposite. It argues from a specific truth, such as an observation in nature, to a more general truth that is the explanation for all observations of the same sort. Bacon's method lays out a procedure to facilitate inferential hypothesis formation by inductive reasoning, as did others who came after him such as John Dewey (see below). On the other hand, the inferential realization of a conjecture or hypothesis as a general explanation for a natural phenomenon requires something more than a method of procedural steps. In further considering how inductive reasoning works, it is important that we examine the spontaneous nature of insight, a phenomenon that may add a missing factor to what Bacon and others attempted to explain.

A brief passage from "The Book of Wisdom", an ancient Biblical epistemological text, alludes to some key insights that will set the stage:

> Wisdom is bright and does not grow dim.
> By those who love Her, She is readily seen,
> And found by those who look for Her.
> Quick to anticipate those who desire Her,
> She makes Herself known to them.
> Watch for Her early and you will have no trouble;
> You will find Her sitting at your gates.
> Even to think about Her is understanding fully grown;
> Be on the alert for Her and anxiety will quickly leave you.
> She Herself walks about looking for those who are worthy of Her and graciously shows
>   Herself to them as they go,
> In every thought of theirs, coming to meet them." (Wisdom 6: 12–17)

The Book of Wisdom was most likely written during the Jewish Hellenistic period, which encompasses a timeframe that includes the first and second century BCE. The author of the text was well versed in the popular philosophical, religious, and ethical writings adopted by Hellenistic Alexandria. It is generally believed that an Alexandrian Jewish philosopher was the author. Philo has often been mentioned in this regard, but this attribution is uncertain. This passage describes *Wisdom* in a female persona as Muse to the human mind, and claims that the love of wisdom and the diligent pursuit of it reduces anxiety and opens the mind to the wisdom already present in the *spontaneous insight* of our thoughts. The idea that Wisdom *graciously shows Herself* to those who are *worthy of Her, in every thought of theirs coming to meet them* is suggestive of a spontaneous unconscious process. This idea is a profound and key insight into the working of the inquiring human mind that has been validated by the work of many philosophers, mathematicians and scientists, since the unknown author of the cited text first wrote about it approximately 2000 years ago.

We know surprisingly little about the cognitive mechanisms that generate insight. The following descriptions of its spontaneous emergence in apes and humans after a period of confusion and frustration, however, begins to suggest that what is called inductive reasoning involves processes that are mediated by the unconscious mind. Virtually every practicing scientist can provide anecdotal evidence from his or her own experience to support the notion that moments of insight often appear "out of nowhere" when they are thinking about, or doing, apparently unrelated things. In fact, most people of all stripes can similarly affirm that they have experienced surprising moments of spontaneous insight. The most recognizable and common manifestation of this phenomenon is the so-called *word on the tip of the tongue*. Everyone at one time or another has trouble remembering the name of a book or its author's name, only to remember it once they have gone on to another topic, whereupon the name is blurted out in the middle of some other conversation, "as if out of nowhere". This supports the idea that, in addition to being spontaneous, insight arises in the form of an inference formed in the unconscious mind while the conscious mind is preoccupied with other matters.

The examination of problem solving in the Great Apes helps to provide an understanding of the unconscious and spontaneous nature of human insight. In 1925 the Gestalt psychologist, Wolfgang Kohler, reported upon a fascinating behavior exhibited by the chimpanzee, Sultan (Kohler W 1926). The experiment provides a strong behavioral demonstration that insight is spontaneous. Kohler gave Sultan two hollow bamboo sticks that Sultan was accustomed to fitting together in play. Kohler then placed some bananas just beyond the point that Sultan could reach using the longer of the two sticks. Sultan then set about trying to get the banana using his arm, then each of the sticks in turn. Sultan even pushed one stick toward the bananas with the other stick and was, thus, able to touch the fruit but not to retrieve it. This maneuver demonstrated that Sultan understood the additive aspect of linear stick extension. What he failed to take advantage of at this point was his understanding, developed in prior play with the sticks, of the importance and utility of *connecting* the two sticks as opposed to merely placing them into contact with each other. After failing to retrieve the banana Sultan retreated to the other side of the cage with his back to the bananas. He was sitting on the box and playing with the sticks in his usual manner, when he appeared to accidentally align the sticks in a straight line. He then quickly fit the end of one stick into the other, as was his custom in play, and promptly used the extended unit to retrieve the bananas. In his account of this episode, Kohler emphasized that the moment of apparent insight came during the playful alignment of the sticks after Sultan's formal attempts to retrieve the banana had been abandoned. Notice that the alignment of the sticks, when Sultan pushed one stick with the other toward the bananas, was not sufficient for the cognitive connection to be made. Insight was achieved only when the sticks were aligned in a context in which they had previously been joined together. This suggests that neural representations retrieved from memory were associated or compared with the perceptual representation of the immediate problem situation to precipitate insight in an apparently spontaneous manner. This view is consistent with work on "the prepared mind" perspective of human cognitive insight (Seifert CM et al. 1995). Perhaps many readers will recognize this pattern in which frustration with a problem and failure to solve it leads to some other distraction that is followed some time later by a moment of spontaneous insight that appears to come out of nowhere. The likely homology between ape and human insight is supported by the striking similarities of the human examples given below to what Kohler reported for Sultan.

It is widely understood that the history of science abounds with examples of the role of serendipity in the achievement of scientific insight. Especially significant in this regard is the inquiry by the mathematician, Jacques Hadamard, into the workings of the mathematical mind (Hadamard J 1945), in which he presents many examples from his own discoveries as well as those of other famous mathematicians. One of these will serve to illustrate the point. Hadamard quotes from Gauss, the famous Nineteenth Century German mathematician, in relation to a theorem of arithmetic that he had tried unsuccessfully to prove for years:

> Finally, two days ago, *I succeeded, not on account of my painful efforts, but by the grace of God. Like a sudden flash of lightening, the riddle happened to be solved.* I myself cannot say what was the conducting thread which connected what I previously knew with what made my success possible. [Emphasis added]

Gauss is clearly unable to explain the origin of his insight. In fact, he disavows any connection to it. He describes it as a "sudden flash of lightening" that comes "by the grace of God". Finally, he states that he cannot explain how he was able to reach the solution from what he "previously knew". Gauss effectively acknowledges in these comments that he was unable to describe a path of deductive reasoning that connected what he previously knew to the solution he finally reached after years of effort. This is a remarkable statement from one of the great European luminaries of mathematics, but it is only one of many that Hadamard presents in support of his thesis, which involves unconscious work on the problem followed by conscious awareness of a solution as a sudden flash of insight.

Another example in the field of mathematics relates the spontaneous insight Irish mathematician, William Rowan Hamilton, experienced in his work on complex numbers. As recounted by mathematical physicists, John Baez and John Huerta (Baez JC and Huerta J 2011):

Hamilton was searching for a three-dimensional number system in which he could add, subtract, multiply and divide. Division is the hard part: a number system where we can divide is called a division algebra. . . . . . To succeed, Hamilton had to change the rules of the game.

Hamilton himself figured out a solution on October 16, 1843. He was walking with his wife along the Royal Canal to a meeting of the Royal Irish Academy in Dublin *when he had a sudden revelation.* In three dimensions, rotations, stretching, and shrinking could not be described with just three numbers. He needed a fourth number, thereby generating a four-dimensional set called quaternions that take the form a + bi + cj + dk. Here the numbers i, j, and k are three different square roots of −1. [Emphasis added]

The discovery of the chemical structure of the benzene ring offers an interesting example of the inquirer's observation of his own musings during what seems to have been a hypnogogic mental state that led to his moment of insight (Rothenberg A 1995):

There I sat, writing on my textbook; but it wasn't going right; my mind was on other things. I turned the chair to face the fireplace and slipped into a languorous state. Again, atoms fluttered before my eyes. Smaller groups stayed mostly in the background this time. My mind's eye, sharpened by repeated visions of this sort, now distinguished larger figures in manifold shapes. Long rows, frequently linked more densely; everything in motion, winding and turning like snakes. And lo, what was that? One of the snakes grabbed its own tail and the image whirled mockingly before my eyes. I came to my senses as though struck by lightning; this time, too, I spent the rest of the night working out the results of my hypothesis. Let us learn to muse, gentlemen, then perhaps we will discover the truth:

"A man not lost in thought
Is given what he's sought,
He'll have it with no effort."
but let us guard against publishing our musings before they have been tested by a vigilant mind.[2]

It would appear that the unconscious mind did so well what Freud originally described. It presented the conscious mind with a representation of hypnogogic cognitive processes in the form of symbolic imagery.[3] The meaning was not lost

---

[2] As translated from Kekule's original German and recounted by Albert Rothenberg in the reference cited in the text.

[3] Kekule's account of his reverie that led to the discovery of the structure of the benzene ring was given to a meeting of chemists. It was written down by the chemist decades after the event. In his careful assessment of this and other translations of Kekule's account Rothenberg (Ibid) argues that the evidence indicates not a dreaming state during sleep but a hypnogogic state that is transitional between wakefulness and sleep. On this basis, he denies participation of the unconscious mind in Kekule's insight. Dreaming during sleep is not a prerequisite for manifestations of unconscious mental activity, however. In any case, it is known that hypnogogic mental cognitions include a process called *autosymbolism*, in which abstract ideas or concepts can be represented as vivid imagery. The imagery may be perceived by the hypnogogic as a symbolic instantiation of more abstract ideas. Moreover, unconscious mental activity can influence the conscious mind even during a state of full wakefulness, as Freud demonstrated so thoroughly in his analyses of wit and slips of the tongue.

on Kekule during his hypnogogic reverie. He was aroused from his mental wanderings as if "struck by lightning".

The common theme in these examples of insight from Gauss, Hamilton and Kekule, as well as many others not cited here, is the sudden spontaneous and sometimes shocking appearance of a solution to a problem after a period of fruitless effort that is followed by a latent period of incubation for a period of time. The supposition is that during this period of incubation, unconscious work is being done until such time that a solution erupts into consciousness.

Finally, the complete quote from a letter of Mozart, which was considered in part in Chap. 4, is also of interest:

> ...thoughts crowd into my mind as easily as you could wish. *Whence and how do they come? I do not know and I have nothing to do with it.* Those which please me, I keep in my head and hum them. . . . . . Once I have my theme, another melody comes, linking itself to the first one in accordance with the needs of the composition as a whole: the counter-point, the part of each instrument, and all these melodic fragments at last produce the entire work. Then my soul is on fire with inspiration...until I have the entire composition finished in my head. . . . . .then my mind seizes it as a glance of my eye a beautiful picture. . . . . .it does not come to me successively..., but it is in its entirety that my imagination lets me hear it. [Emphasis added]

As with Gauss, Hamilton and others, Mozart did not know from where his inspiration came, and he explicitly states that he had nothing to do with it! All he can say is that his *soul is on fire with inspiration.* What is especially interesting about this description by Mozart is that he appears to have been in touch with unconscious cognitive mechanisms while he was awake, or possibly in a hypnogogic state similar to that of Kekule. Another aspect worthy of special note in Mozart's description of his creative process is how he was able to experience or hear it in his mind not successively but entirely in one moment of time. This is indicative of a state in which the functions of the right cerebral cortex may have dominated Mozart's awareness. We will see below that this is consistent with right cerebral hemispheric mechanisms of parallel information processing as opposed to the sequential information processing of the left hemisphere.

This picture of spontaneous insight, as a contributing factor in problem solving, stands in stark contrast to the one painted by the rational empiricists of the 17th, 18th and 19th Centuries. Philosophers like Francis Bacon created a framework which posited that a *procedure* of inductive reasoning could help to formulate hypotheses that could be experimentally tested to discover truth about the natural world. Yet, neither Hadamard, nor Gauss, nor Hamilton, nor Mozart, could explain their great insights in terms of antecedent procedural or logical steps; and this experience is quite common among inquirers of all types. Do the philosophers have anything new to say about the nature of inquiry since the time of Bacon? In fact, they do and of special interest to us are the American philosophers John Dewey and Arthur Bentley. We will see that, although Bacon and Dewey-Bentley showed that a procedure for inductive reasoning can set the stage for discovery, a vital and conclusive step is provided by the phenomenon of spontaneous insight.

Epistemology – The Philosophy of Inquiry

Epistemology is sometimes referred to as the theory of knowledge[4] as it has been expounded by philosophers from antiquity until the more recent work of the American philosophers John Dewey and Arthur Bentley. A quote from one of Dewey's later essays, "Time and Individuality" (Dewey J 1940) is helpful at the outset of our consideration of his work in epistemology:

> ...classic philosophy maintained that change, and consequently time, are marks of inferior reality, holding that true and ultimate reality is immutable and eternal. Human reasons, all too human, have given birth to the idea that over and beyond the lower realm of things that shift like the sands on the seashore there is the kingdom of the unchanging, of the complete, the perfect. The grounds for the belief are couched in the technical language of philosophy, but the grounds for the cause is the heart's desire for surcease from change, struggle, and uncertainty. The eternal and immutable is the consummation of mortal man's quest for certainty.

When he says, "the eternal and immutable is the consummation of mortal man's quest for certainty", Dewey is clearly saying that the cause for human belief in an eternal immutable realm lies in humanity's desire to escape uncertainty. Dewey alludes here to two of the major themes in his epistemology. One theme concerns the futility and fallacy of any quest for certainly, which has as its corollary the concept that the demonstration of empirical truth is always imperfect and only accomplished by gradual iterative steps. The second theme relates to Dewey's rejection of the existence of a *true and ultimate reality* that is *immutable and eternal*, such as Plato's *realm of ideals or perfect forms*. Dewey's development of this idea in his epistemology led him and Arthur Bentley to propose that knowledge exists in mind only, that it has no ontological validity apart from knowing mind; and, therefore, that the only valid epistemological dynamic is the dyadic one between a knowing mind and the object known.

In regard to the quest for certainty, we begin by considering a description of the course of human inquiry that is consistent with Dewey's epistemology from the work of Rollo Handy and E.C. Harwood (Handy R and Harwood EC 1973). It speaks of the confusion of the scientist in the face of the unknown, and of an iterative process of observation, hypothesis formation and further observation:

> In the course of inquiry, as the inquirer initially observes and measures he may note connections among the things measured and imagine other possible connections...Partial tentative descriptions are developed from initial observations. The inquirer, temporarily baffled in his effort to achieve useful adequate description, imagines various possibilities or notions among which he selects the seemingly more promising as the guide to further observation. If lucky or skillful or both, his additional observations develop further the original tentative descriptions. If not adequate for the problem confronted by the inquirer, he again is baffled and develops notions or imagines other possible connections among things, from which he chooses a guide for further observations and so on, until satisfactory

---

[4]Ironically Dewey argued that *knowledge*, per se, does not exist apart from a knowing mind! This has already been discussed in Chap. 1. We will examine the matter in greater detail here.

description is achieved. In the course of these procedures of inquiry, the inquirer may explore many blind alleys, discard many of his observations, and begin over again at various stages, perhaps many times...the sequence of proceeding is not from elaborately formulated hypotheses to testing of them by subsequent observation of facts. Rather the sequence in successful inquiry seems always to be from observation and measurement of initially selected aspects and phases of the problem situation to partial, tentative inadequate descriptions, followed by conjectures about possible but as yet unobserved connections, which in turn require new observations, etc.

It is easy to see, from this accurate (though not often acknowledged) description of the tentative and confusing course of scientific inquiry, why John Dewey and Arthur Bentley maintained in "Knowing and the Known" (Ibid) that science proceeds erratically in a pattern of fits and starts, as the inquirer proceeds ever so slowly to cast away uncertainty in making the transition from confusion to *knowing*. Moreover, according to Dewey and Bentley, *knowing* is always relative and incomplete. For them, truth is apprehended incrementally and always imperfectly. They resoundingly reject the quest for certainty as futile because the results of inquiry are always incomplete and subject to modification to achieve a deeper understanding of the phenomenon under investigation. On this, Rollo Handy states elsewhere (Handy R 1973):

Many of the procedures of inquiry rejected by Dewey and Bentley involve what Dewey (1929) called the "quest for certainty", which was characteristic of most work of the ancient Greek philosophers and often found today, if in a more obscure form. The difficulties, hazards, and uncertainties of life are so frustrating that often humans long for an absolute certainty. That absolute certainty may be thought to reside in a knowledge of a Platonic heaven of ideas, of Aristototelian essences, of supernatural existences, of epistemological "incorrigibles", or of logical certainties, but some absolutely assured way of apprehending important truths allegedly is available...

Dewey and Bentley rejected certainty as the objective of inquiry and emphasized forcefully that all data, all facts, and all interpretations of facts are subject to modification and possible rejection as inquiry proceeds. What is known is not a terminus outside or beyond inquiry, but is a goal within inquiry. In contrast to conventional views, the Dewey-Bentley procedures take human knowings as observable behaviors; man's most complex inquiries are themselves to be inquired into in the same general way that any scientific subject is investigated.

Handy's and Harwood's description of the Dewey-Bentley view of the chaotic and often confusing nature of scientific inquiry, until a working hypothesis is formed, does not refer explicitly to the cognitive mechanisms that are responsible for hypothesis formation. In the Dewey-Bentley construct of epistemology, one is left to wonder how, in the confusion that reigns throughout the process of inquiry, clarity often emerges suddenly at a moment of insight to provide a tentative hypothesis that may explain a phenomenon under investigation. In their account of the course of inquiry, Handy and Harwood describe a progression from data to conjecture. This is the transition of interest, wherein confusion gives way to clarity, and wherein one may postulate that an unconscious process of inductive reasoning gives birth to spontaneous insight in the form of a conjecture or hypothesis. This is followed by experiment that incorporates elements of deductive reasoning in the design of the experiment; the control groups versus experimental groups and so forth. Therein we have the aforementioned interplay of inductive and deductive reasoning in inquiry. Here is what Dewey has to say about this interplay in his book, "How We Think" (Dewey J 1910):

> While induction moves from fragmentary details (or particulars) to a connected view of a situation (universal), deduction begins with the latter and works back again to particulars, connecting them and binding them together. The inductive movement is toward *discovery* of a binding principle; the deductive toward its testing, confirming, refuting, modifying it on the basis of its capacity to interpret isolated details into a unified experience. So far as we conduct each of these processes in the light of the other, we get valid discovery or verified critical thinking. [Emphasis added]

We should take note here of the "binding" together of deductively demonstrated truths! Concerning the control, or regulation, of inductive reasoning, Dewey continues (Ibid, p 84):

> Control of the formation of suggestion is necessarily indirect, not direct; imperfect, not perfect. Just because all discovery, all apprehension involving thought of the new, goes from the known, the present, to the unknown and absent, no rules can be stated that will guarantee correct inference. Just what is suggested to a person in a given situation depends upon his native constitution (his originality, his genius), temperament, the prevalent direction of his interests, his early environment, the general tenor of his past experiences, his special training, the things that have recently occupied him continuously or vividly, and so on; to some extent even upon an accidental conjunction of present circumstances. These matters, so far as they lie in the past or in external conditions, clearly escape regulation. A suggestion simply does or does not occur; this or that -suggestion just happens, occurs, springs up. If, however, prior experience and training have developed an attitude of patience in a condition of doubt, a capacity for suspended judgment, and a liking for inquiry, indirect control of the course of suggestions is possible.

Dewey is quite clear here that "no rules can be stated" for why or how, "a suggestion simply does or does not occur; this or that -suggestion just happens, occurs, springs up." On the other hand, he quickly follows with, "If, however, prior experience and training have developed an attitude of patience in a condition of doubt, a capacity for suspended judgment, and a liking for inquiry, indirect control of the course of suggestions is possible." This seems to contradict his prior statement about not knowing how a "suggestion just happens"; but Dewey is making an important point about cultivating a sense of familiarity and comfort with ambiguity, that is inherent in the state of uncertainty, as a necessary pre-condition for the emergence of spontaneous insight. He goes even further to state, as Bacon before him, that despite the aforesaid lack of rules and the spontaneous nature of insight, it is possible to regulate the processes of observing and amassing data to facilitate the inferential leap that induction provides (Ibid, p. 86):

> Scientific induction means, in short, all the processes by which the observing and amassing of data are regulated with a view to facilitating the formation of explanatory conceptions and theories. These devices are all directed toward selecting the precise facts to which weight and significance shall attach in forming suggestions or ideas. Specifically, this selective determination involves devices of (I) elimination by analysis of what is likely to be misleading

and irrelevant, (2) emphasis of the important by collection and comparison of cases, (3) deliberate construction of data by experimental variation.[5]

Dewey clearly states that hypothesis formation by inductive reasoning is spontaneous; but, like Francis Bacon before him, he also lays out a proper context and procedure for the formation of what he calls *scientific* induction. Even so, the cognitive mechanisms of induction remain unspecified by Dewey in "How We Think". He steadfastly refuses to appeal to the operation of the unconscious mind, most likely because of his insistence upon an *observable* behavioral paradigm for scientific inquiry. In spite of Dewey's insistence on empirical rigor by focusing on *inquiry as behavior*, there is ample basis for giving due consideration to a role for the unconscious mind as the locus of at least some of the cognitive mechanisms of inductive reasoning and as the source of spontaneous insight. After all, even in Dewey's own time the work Sigmund Freud, who was famous for demonstrating the role of the unconscious mind in dreaming, wit, and spontaneous slips of the tongue, was available to point in this direction (Freud S 1995). Finally, in 1945 during the latter part of Dewey's life there was also the previously mentioned work of Hadamard, which provided an abundant record of anecdotal evidence that supported the role of the unconscious mind in the formation of insight.

On the other hand, and at least partially in his defense, Dewey did not have the benefit of modern advances in cognitive neuroscience that derive from experimental work on regional mapping of brain activity during various mental tasks. Neuroimaging studies (Mark JB et al. 2004) have shown that solution of a semantic problem such as an anagram by spontaneous insight, as opposed to a process of deductive reasoning, is associated with increased functional MRI (fMRI) activity in the right anterior temporal lobe. Moreover, a sudden burst of high-frequency EEG activity was noted in the same area about 300 ms before the subjects indicated that they had reached a solution by insight. These effects were not observed for non-insight solutions. Enhancement of the regional fMRI signal has been interpreted as the neurological concomitant of unconscious processing of diffusely related semantic information by the right hemisphere until a time that a solution is achieved, at which point the high frequency EEG activity is associated with the emergence of the solution to conscious awareness. Interpreting the results of this fMRI study, the authors suggested that it is not surprising to find increased activity localized to the right hemisphere, because it is uniquely constituted to efficiently seek associations of the altered sequence of letters in the anagram with "distally related" semantic content that provides the letter sequence required for the solution. The left hemisphere, on the other hand, is more apt to locate strong connections to closely related but irrelevant semantic content that might actually block a solution. Distraction from the problem situation, as in the anecdotal stories of sudden insight given by Hadamard and others, may allow the weaker connections to distally related semantic content to become stronger after which at some threshold, the solution emerges into consciousness. It is tempting to see, in the different information processing

---

[5]Note the similarity to the method that Bacon proposed.

modalities of the left and right cerebral cortices, computational mechanisms that are best suited for deductive and inductive reasoning, respectively. We have seen that deductive reasoning is most concerned with justifying the sequential steps in a syllogistic argument that proceeds from a general truth to a specific instance. The sequential processing mechanisms of the left hemisphere are best suited for this task; but inductive reasoning must search for major inferential leaps to "distant" related cognitive constructs that may provide a general solution to a specific instance of a problem. The parallel processing mechanisms of the right hemisphere, in turn, may be best suited for this task, which proceeds not by deductive argument but according to the relationships among elements of a problem and the elements in long-term memory that may be critical to defining a leap of inference. In fact, we should consider whether, in cases of spontaneous insight, an inferential solution to a problem is computed by unconscious cognitive algorithms in the right hemisphere that are similar to what both Bacon and Dewey recommend as conscious cognitive procedures or listing of examples, counter-examples, and degrees of a phenomenon to facilitate the formulation of the general principle of which the phenomenon being studied is a special case. A difference between these two different methods of formulating a conjecture may be that in the case of unconscious spontaneous insight a period of distraction is required, whereas in the method of *scientific induction* conscious procedures might facilitate the inferential leap without a period of distraction.

The incremental nature of the inquiry's progression from confusion to under-standing, especially empiricism's inability to demonstrate certain truth, is a vital part of the epistemology of Dewey and Bentley. Equally important, however, is what they say about the second major theme of "Knowing and the Known": the status of *knowledge, per se*. Dewey and Bentley take issue with the work of other philoso-phers and logicians who imbue knowledge with any existence apart from the *act of knowing*. Instead they present a transactional paradigm in which there is only *knowing mind* and the *object of its knowing*. They even take issue with and deplore the use of the word *knowledge* as being illogical and confusing. It may be difficult to eliminate use of the word knowledge as the noun that specifies *that which is known*, but even this definition makes clear that in using *knowledge* in the passive voice, the essential *knower* is relegated to a merely implicit status. Certainly, *that which is known* requires a *knower*. For Dewey and Bentley, understanding something means *to know* it in the active sense. *Knowing* as a *cognitive process* is paramount for them, and use of the word *knowledge* should be avoided.

At the outset of "Knowing and the Known", Dewey and Bentley comment in regard to C. S. Peirce, a former teacher of Dewey:

> Consider again what Peirce, cutting still more deeply, wrote about the sign, "lithium" in its scientific use: "The peculiarity of this definition – or rather this precept that is more serviceable than a definition – is that it tells you what the word 'lithium' denotes by prescribing what you are to do in order to gain a perceptual acquaintance with the object of the word".

Notice the "perceptual"; notice the "object" of the "word". There is nothing here that implies a pattern of two orders or realms brought into connection by a third intervening thing or sign. This is the real Peirce; Peirce on the advance – not bedded down in the ancient swamp."

Dewey and Bentley continue:

The cosmic pattern we shall employ, and by the aid of which we shall make our tests, differs sharply from the current conventional one and is in line with what Peirce persistently sought. It will treat the talking and talk-products or effects of man (the namings, thinkings, arguings, reasonings, etc.) as the men themselves in action, not as some third type of entity to be inserted between the men and the things they deal with. To this extent it will not be three-realm, but two realm: men and things. . .It rests in the simplest, most direct matter-of-fact, everyday, common sense observation. Talking organisms and things – there they are: if there, let us study them as they come: the men talking.

We may well ask, "isn't *knowledge* recorded in the volumes found in the libraries of the world". A strict interpretation of Dewey's and Bentley's philosophy requires that we answer that what is recorded there are symbols on a page. They are known only when they are written and when they are read. *There is nothing known without the mind that knows it.* The application of this aspect of Dewey's and Bentley's philosophy to the notion of Consilient Truth will provide a key step in a modified proof from truth for the existence of God. Ironically, Dewey was a humanist[6] and a confirmed atheist, as indicated in much of his writing as here in "Teacher Magazine"[7] (Dewey J 1933):

There is no god and there is no soul. Hence, there is no need for the props of traditional religion. With dogma and creed excluded, then immutable truth is dead and buried. There is no room for fixed law or permanent moral absolutes.

We can find the basis for Dewey's atheistic conviction in his strict adherence to an empirical approach in his work in philosophy, psychology and education, which would put him squarely in the camp of those scientist-philosophers who object to the existence of God as an unnecessary being in an unobservable transcendental realm. Yet even now, before advancing to the culminating argument of this book, we can see this position weakening at the foundation. For it is the scientists and philosophers themselves who are seeking a transcendental cause of our vast universe, which may be only a miniscule part of an eternal multiverse that some consider to be created as a "virtual reality" on a cosmic quantum computer by and for the pleasure of a "transcendent being" who dwells in the "true reality" beyond this "virtual" one.

---

[6]He was in fact an author of the *Humanist Manifesto*.

[7]This citation is given in a number of other sources, but I have not found the original.

A Thought Experiment

As explained above, John Dewey advocated for an epistemology that is focused on the inquirer *in action*. This is totally consistent with his interest not only in philosophy, but also psychology. We can see that Dewey's epistemology was informed by his interest and insights in both areas of inquiry. His focus, with Arthur Bentley in "Knowing and the Known", is on a dyadic "cosmic pattern" that takes account of the act of knowing and the object known. As we have seen, their epistemology rejects any intermediate third realm or agency, such as knowledge, per se, or any transcendental realm in which knowledge exists. We see this very clearly in this edited version of the quote that was presented above:

> The cosmic pattern we shall employ,... will treat the talking and talk-products or effects of man (the namings, thinkings, arguings, reasonings, etc.) as the men themselves in action, not as some third type of entity to be inserted between the men and the things they deal with. To this extent it will not be three-realm, but two realm: men and things... Talking organisms and things – there they are: if there, let us study them as they come: the men talking.

How do those "men talking", Dewey's inquirers, come to know what they think they know about the objects of their knowing? Dewey was essentially advocating for a *behavioral* orientation to the philosophical study of inquiry, and he was right to enunciate such a seminal idea. We stand a better chance, however, of illuminating the inquirer's transition from confusion to insight if we take Dewey's behavioral orientation to epistemology to an imaginary empirical fulfillment in a behavioral experimental paradigm. Behaviorism rose up as a school of psychology in reaction against the idea that *thoughts*, *perceptions*, and various other *mentations* could be proper objects of scientific investigation owing to their inherently subjective nature. How does one observe an idea or thought?[8] For this reason J.B. Watson and others made a strong case for the study of behavior as an objective observable which would allow for a rigorous scientific approach to phenomena mediated by the brain or mind. We can design an empirical behavioral approach to the study of inquiry; and in doing so create a union between Dewey's interests in Philosophy and Psychology.

Imagine that you are recruited into an experiment that was advertised as a study of eye movements. At the appointed time, you arrive at the laboratory and are seated in a comfortable chair with a screen a few feet in front of you. You are told that a black dot will appear briefly in the middle of the screen, and that it will immediately be followed by an image of the type shown in Fig. 7.1 A or B.

The scientist running the experiment also instructs you that your eye movements will be tracked by instrumentation and that a record of these movements will be kept. The sequential 10 s presentations of the images begin. Imagine now that you, as a subject in this thought experiment, notice that the smiling and crying baby face composite images seem to appear in a random order. You also notice that the geometric shapes vary in position as well as shape. The same figures are not always

---

[8] According to Theory of Mind, we make inferences all the time about what other people are thinking and feeling, but these inferences are hypotheses not direct observations.

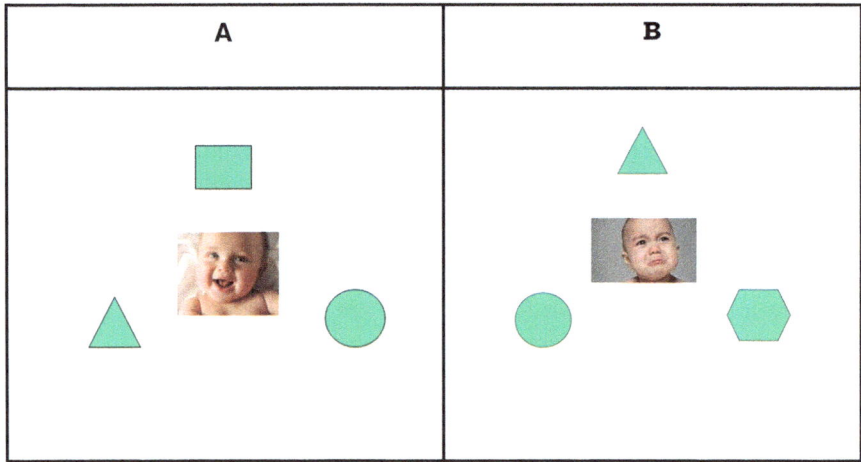

Fig. 7.1 Panel A and B show the images that each include a baby face among three geometric shapes. Each of these will be shown for 10 s. Each image will be presented for 10 s, after which the black dot in the middle of the screen will reappear

present. Most likely, you would be confused about what all this means. All you would know at this point in the experiment would be that your eye movements are being studied. Imagine now that you, the subject, keep looking as you have been instructed to do at the visual scenes as they are presented. What does the experimenter see on his screen? The same images appear on the experimenter's computer screen as on the screen you are viewing, but you do not see what she sees. This is sown in Fig. 7.2.

As the random sequence of images progresses, the experimenter will look for changes in the pattern of eye movements to determine if any of the geometric shapes begin to provoke greater interest than would be expected by chance. A pattern like the one in Fig. 7.3 would indicate that the square shape had acquired some special significance.

This pattern might be quantified as number of repeat gazes toward the square during each 10 s image presentation. Other measurements would also be indicative of an acquired significance of the square. Some of these indicators would be decreasing latency to the first gaze toward the square, and/or increased duration of gaze on the square as image presentations progress; and increasing probability that the first directed eye movement is toward the square as image presentations progress, etc.

How might subjects react in a real experiment like this one? They would begin with the understanding given in the instructions, that this is an experimental study of eye movements. As the experiment progressed, however, they would have uncertainty about the unfolding events, and they might wonder why eye movements would be studied in this particular way. At some point, some of the subjects might become consciously aware that the square was only present when the smiling baby

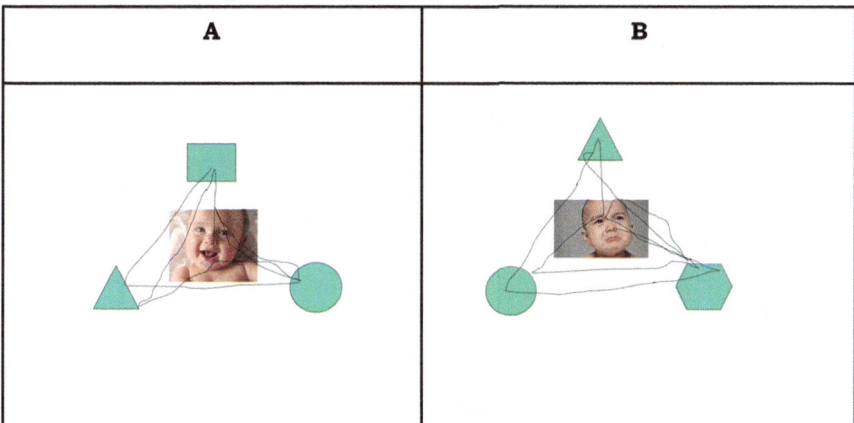

Fig. 7.2 Real-time tracings of your eye movements appear on the experimenter's screen during each 10 s image presentation. The eye movements are traced as thin black lines over the image and recorded electronically for later analysis. (**a**). Eye movement tracing for happy baby image. (**b**) Eye movement tracings for crying baby image

Fig. 7.3 Preferential gazing toward the square, perhaps alternating with gazing at the smiling baby face suggests that the square has acquired a greater ability to command the subject's attention than expected on the basis of chance

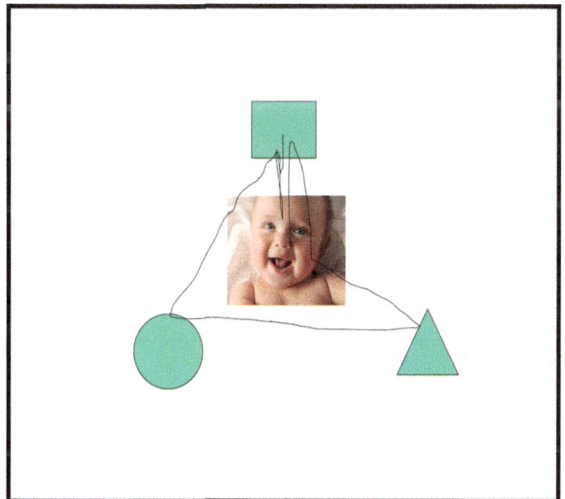

face was also present. This would be a correct inference on the part of the subjects. The experiment is designed to study the classical conditioning of eye movements, but ultimately reveals much more. There is a one hundred percent correlation between the presence of the smiling baby face in the center of the image and the square being one of the three shapes shown. Figure 7.4 illustrates the relationships in this experiment. The smiling baby face is the unconditional stimulus. This derives from the inherent ability of a smiling baby to evoke a positive instinctive emotional response in adult humans such as gazing and smiling back at it. The conditional

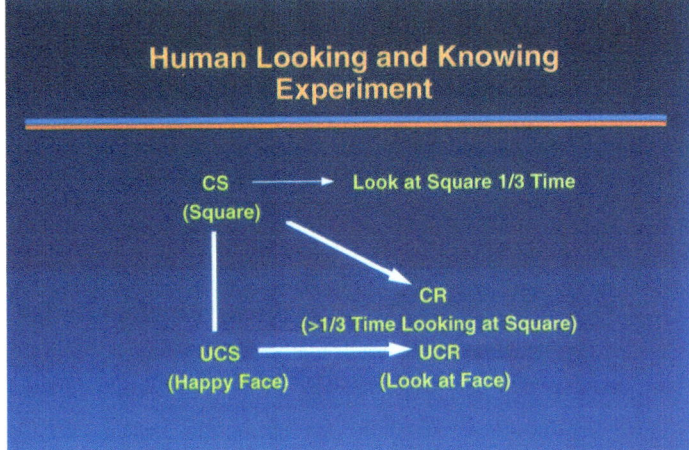

Fig. 7.4 Diagram that illustrates the contingent relationship between the UCS (happy baby face) and the CS (square). The UCS elicits looking behavior toward the baby face, while initially the square elicits no special interest as indicated by the equal allocation of time looking at the three shapes (1/3 each). After repeated pairing between the CS and UCS, however, the square elicits greater interest which is now indicated by disproportionately more time spent looking at it (>1/3)

stimulus is the square, which is one of six neutral geometric shapes (square, circle, triangle, hexagon, diamond and star) any three of which are selected randomly for each trial, except that in all cases the square appears with the smiling baby face. Initially we would expect the square to be equipotential with the other shapes in commanding the subject's attention. This equivalence among the shapes might be indicated by an equal amount of time looking at each of them, or an equal number of eye movements toward each, etc. After repeated pairings, however, the experimenter would notice a change in the gazing behavior of the subjects. The square would begin to command increased attention owing to its pairing with the smiling baby face. The square would have acquired new meaning, by virtue of which, it would provoke sign-tracking just as we saw in the pigeon in Fig. 6.3 of Chap. 6. In this experiment, sign-tracking would manifest as a change in eye movement that is consistent with the emergence of a new understanding about the significance or meaning of the square.

If you are beginning to suspect that there is more to this experiment than "meets the eye", you are right. It is an experiment about human looking, knowing, and objects known. It goes to the heart of Dewey's philosophy. This classical conditioning experiment is designed to study the behavior of an inquirer as a transition is made from uncertainty to insight. The subject in the experiment is actually meant to represent an inquirer confronted with a problem, such as a scientist who is initially confused by a phenomenon or data that he or she is trying to understand. In our thought experiment, the subject stands in for, or models, the scientist/inquirer by virtue of being put into a situation that is poorly understood initially.

Humans, indeed all organisms, are confronted constantly by uncertainty about the meaning of events or phenomena that unfold around them. The inquirer (scientist, philosopher or anyone trying to understand something) is initially confused but as he or she continues to observe the world, relationships among stimuli are detected and tracked by the nervous system. We are told in the study of statistics that correlation is not causation, and this is certainly true, but it *absolutely is* information that captures the selective attention mechanism of the inquirer's brain and directs it toward informative relationships among events that have the potential to provoke spontaneous insight.[9] Eventually insight about the meaning of events reaches the conscious mind from its unconscious wellspring. In the case of our thought experiment, the insight would be the realization that *the square and smiling baby face occur together.* Classical conditioning is the Muse that whispers the revelations of spontaneous insight to the inquiring human mind. Moreover, it is the Muse that now reveals to us a vital clue about the neurological mechanism of inductive reasoning. In the case of classical conditioning, the brain must recognize that two specific stimuli, the CS and UCS, share a feature in common, namely that they are proximate to each other in time and space. They occur repeatedly in spatiotemporal correlation. This instance of spontaneous insight is the realization of vital information. It provides the solution to the subject's question in the experiment: "what is going on here?". Recall that the result of inductive reasoning is the spontaneous realization of a general principle from specific instances or observations that fall under the category of the general principle. In our human looking and knowing experiment, two stimuli are perceived initially as separate distinct objects, but after a number of pairings the subject's brain detects and recognizes something that equates them. They always occur together. They are bound in time and space and the realization, or detection, of that information provides a new way to understand what was not understood before. The subject has figured out what the experiment is about. Something very much akin to this happens in the unconscious mind whenever a general principle is realized based on observations of specific instances of that principle. The unconscious mind is always searching for what binds specific observations into a potential understanding of new meaning about what has been observed: the overarching principle that explains the specific instances that have been observed. So, perhaps the neural mechanisms of classical conditioning provide a simple case or model for how the brain accomplishes the amazing phenomenon of spontaneous inferential insight based on observed events.

Schrödinger was right. Organisms must detect, approach, and assimilate information, or order, to ensure survival. The information inherent in correlated stimuli is so vital to survival that the ability of nervous systems to accomplish this task was present in the first multicellular organisms. The cell and molecular neurobiology of classical conditioning is not only ancient, but also highly conserved. It provides the basis of synaptic plasticity and memory that allows organisms to adaptively modify

---

[9]Spontaneous insight is the meaning that the inquirer's brain infers from correlations detected among salient stimuli in the perceptual field or in Long-Term Memory.

behavior based on the detection of information that is inherent in correlated events. It is amazing and humbling to contemplate that the neurobiology of our loftiest insights is founded upon signaling mechanisms that were present in some of the earliest single-celled organisms, which in turn provided the basis for classical conditioning in multicellular invertebrates of the Cambrian era. With this background, we can venture a prediction about the behavior of multicellular extra-terrestrial life-forms if they are ever discovered. They will be capable of detecting and tracking correlations among discrete objects in their perceptual field; and their behavior will be modifiable on the basis of the information that is detected in those relationships. In other words, they will be capable of classical conditioning! How can we have confidence in this prediction? *Organisms must detect, recognize, approach and assimilate information[10] to defend against the disorganizing effects of entropy; and entropy will act as the selective pressure that sculpts the genetic mechanisms by which those organisms bequeath to their progeny the necessary metabolic and behavioral mechanisms for doing so!*

References

Baez JC, Huerta J (2011) The strangest numbers in string theory. Sci Am 304(5):60–65

Dewey J (1910) How we think. D.C. Heath, Lexington, MA, p 81

Dewey J (1933), Soul-searching. Teacher Magazine, September 1933, p 33

Dewey J (1940) Time and individuality. In: Dewey J, Boydston JA (eds), The later works, 1925–1953. Southern Illinois University Press. Carbondale, 1988, pp 98–99

Freud S. (1995) The basic writings of Sigmund Freud (Trans. and Ed. AA Brill). The Modern Library, New York

Hadamard J (1945) The psychology of invention in the mathematical field. Princeton University Press. Reprinted in 1954. Dover Publications Inc. Mineola, New York

Handy R (1973) The Dewey-Bentley transactional procedures of inquiry. Psychol Rec 23:305–317

Handy R, Harwood EC (1973) Useful procedures of inquiry. Behavioral Research Council, Great Barrington, p 14

Jürgen K (2015), Francis Bacon. In: Zalta EN (ed) 5. Stanford encyclopedia of philosophy, Summer 2015 edn. http://plato.stanford.edu/archives/sum2015/entries/francis-bacon/

Klein J (2015) Francis Bacon. In: Zalta EN (ed) The Stanford encyclopedia of philosophy, Summer 2015 edn. http://plato.stanford.edu/archives/sum2015/entries/francis-bacon/

Kohler W (1926) The mentality of apes, vol 1976. Liveright, New York, pp 125–128

Livio M (2005) The equation that couldn't be solved. Simon and Schuster, New York

Mark JB et al (2004) Neural activity when people solve problems with insight. PLoS Biol 2:500–510

Rothenberg A (1995) Creative cognitive processes in Kekule's discovery of the structure of the benzene molecule. Am J Psychol 108:419–438

Seifert CM et al (1995) In: Sternberg RJ, Davidson JE (eds) Demystifying cognitive insight: opportunistic assimilation and the prepared mind perspective. MIT Press, Cambridge, pp 65–124

---

[10]Information is assimilated via the agency of bioenergetic and neurobiological mechanisms.

Chapter 8
Consilience, Truth and the Mind of God: A Synthesis

Richard J. Di Rocco and Arthur J. Kyriazis

> *To be or not to be, that is the question.*
>
> Hamlet, Act III, Scene I. William Shakespeare
>
> *The fool has said in his heart, 'There is no God'.*
>
> Psalm 14:1
>
> *Ehyeh-Asher-Ehyeh, I-Will-Be-Who-I-Will-Be.*
>
> Exodus 3:13–14

Abstract As the title implies, this chapter will review the main conclusions of the preceding chapters as a prelude to consideration of the essential ontological question concerning the existence of God. We begin with "The Metaphysical Poem of Parmenides", which in its own right provides an intriguing basis for belief in a necessary Being that is the source of all being. The main philosophical arguments for the existence of God are briefly presented, beginning with St. Augustine's original "Argument from Truth", which is then followed in historical order by Boethius' "Argument for the Necessity of a Supreme Good", and detailed explanation of St. Anselm's "Ontological Argument". The five arguments of St. Thomas are then mentioned with particular attention given to "The Argument of the First Cause" and "The Argument of Contingency", which together lead to the existence of a Necessary Being that is the Self-Sufficient First Cause all that exists. St. Thomas's "Argument of the First Cause" and "Argument of Contingency" are re-evaluated

R. J. Di Rocco (✉)
Psychology Department, University of Pennsylvania, Philadelphia, PA, USA

Psychology Department, St. Joseph's University, Philadelphia, PA, USA
e-mail: richdi@upenn.edu

A. J. Kyriazis
Phila Pharma & Biotech Inc (PPBI), Seven Cities Consulting Group, Philadelphia, PA, USA
e-mail: trebizond1461@gmail.com

© Springer Nature Switzerland AG 2018
R. J. Di Rocco, *Consilience, Truth and the Mind of God*,
https://doi.org/10.1007/978-3-030-01869-6_8

in the context of the existence of an eternal multiverse, in which case we must ask whether such an entity could be the Necessary Being that is the sufficient cause of itself. We then show why the interdependent collection of pocket or bubble universes that comprise the eternal multiverse cannot be the sufficient cause of itself; and why the parallel argument that atheists make against the self-sufficiency of God as the Necessary Being is answered by the *Modified Argument from Truth*, described herein. We show the essential role played by the existence of infinite and eternal Consilient Truth, and the epistemology of Dewy and Bentley which posits that knowledge, per se, has no existence of its own but rather must exist in *mind knowing truth*. This argument leads to the existence of an eternal mind that knows eternal Consilient Truth.

Keywords Metaphysics · Classical and medieval proofs of the existence of god · The ontological argument · Five proofs of Thomas Aquinas · The modified argument from truth · Co-eternal god and multiverse

The Metaphysical Poem of Parmenides

In the preceding chapters, we explored what science can tell us about reality from the moment of the Big Bang, to the creation of life from non-living matter, and finally to the emergence of the modern human mind. We have come full circle in the quest to understand reality. It is time now to turn our evolved faculty of intelligence to the ultimate ontological question. Is God the eternal, transcendent, intelligent Being that is responsible for all that exists, or is the universe its own self-sufficient cause? The early chapters of this book explained why neither science nor religion is competent to prove or disprove the existence of God. Pronouncements on this question by religion and science, therefore, are made on the basis of faith alone. In this chapter, we will bring the logic of philosophical argument to bear on the question of God's existence in the search for ultimate meaning, beginning with the Poem of Parmemides.

The metaphysical poem that was composed in the fifth century BCE in southern Italy by the Greek philosopher, Parmenides, is of great interest in our inquiry at this point. Parmenides anticipates many of the themes we have already discussed, and his thought heavily influenced the Greek and Roman philosophers that succeeded him, and also the medieval philosophers, we will discuss in this chapter. As mentioned in a previous footnote, the poem exists mostly in fragments (fr.) that were preserved because he was quoted by his Greek intellectual descendants. The poem's metaphysical argument is complete, however, as explained below.

According to an entry in "The Stanford Encyclopedia of Philosophy" (Palmer 2016), we know how the relevant portion of Parmenides' Poem comes down to us:

> The Alexandrian Neoplatonist Simplicius (6th c. CE) appears to have possessed a good copy of the work, from which he quoted extensively in his commentaries on Aristotle's Physics and De Caelo. He introduces his lengthy quotation of fr. 8.1–52 as follows: "Even if one

might think it pedantic, I would gladly transcribe in this commentary *the verses of Parmenides on the one being,* which aren't numerous, both as evidence for what I have said and because of the scarcity of Parmenides' treatise." Thanks to Simplicius' lengthy transcription, we appear to have entire Parmenides' major metaphysical argument demonstrating the attributes of "What Is" (to eon) or "true reality" (alêtheia). [Emphasis added]

The portion of Parmenides' Poem that encompasses his metaphysical argument is noteworthy for its prescience regarding the unitary nature of reality, which was supported approximately 2500 years later by quantum physics; and also in regard to the eternal nature of the multiverse proposed by some modern cosmologists. While this latter point is subject to definitive proof that the multiverse is real, it is clear that Parmenides anticipates a question that is of great interest to cosmologists today. There is little doubt that Parmenides' Metaphysical Poem influenced the thinking of the Greek and medieval philosophers that succeeded him on the question of God's existence. The passage quoted below also seems to involve aspects of a perspective from a higher-dimensional reality that may be described as the view from *no-when* and *no-where*. Parmenides also may have been one of the first to write poetry using an expository method that involves the instruction of a mortal by a divine being. This style was emulated by Virgil, Boethius (see below) and more recently by Dante and others. In Parmenides' Poem, a goddess instructs a youth in two ways of thought; the one false and the other true, which Parmenides sets out once he has dispatched the false way of thought. The section that describes the true way of knowing is quoted in full below. The entire passage is worthy of emphasis (Burnet 1920):

One path only is left for us to speak of, namely, that It is. In this path are very many tokens that what is is uncreated and indestructible; for it is complete, immovable, and without end. Nor was it ever, nor will it be; for now it is, all at once, a continuous one. For what kind of origin for it wilt thou look for? In what way and from what source could it have drawn its increase? . . . I shall not let thee say nor think that it came from what is not; for it can neither be thought nor uttered that anything is not. And, if it came from nothing, what need could have made it arise later rather than sooner? Therefore must it either be altogether or be not at all. Nor will the force of truth suffer aught to arise besides itself from that which is not. Wherefore, justice doth not loose her fetters and let anything come into being or pass away, but holds it fast. Our judgment thereon depends on this: "Is it or is it not?" Surely it is adjudged, as it needs must be, that we are to set aside the one way as unthinkable and nameless (for it is no true way), and that the other path is real and true. How, then, can what is be going to be in the future? Or how could it come into being? If it came into being, it is not; nor is it if it is going to be in the future. Thus is becoming extinguished and passing away not to be heard of.

Nor is it divisible, since it is all alike, and there is no more of it in one place than in another, to hinder it from holding together, nor less of it, but everything is full of what is. Wherefore it is wholly continuous; for what is, is in contact with what is. Moreover, it is immovable in the bonds of mighty chains, without beginning and without end; since coming into being and passing away have been driven afar, and true belief has cast them away. It is the same, and it rests in the selfsame place, abiding in itself. And thus it remaineth constant in its place; for hard necessity keeps it in the bonds of the limit that holds it fast on every side. Wherefore it is not permitted to what is to be infinite; for it is in need of nothing; while, if it were infinite, it would stand in need of everything.

The thing that can be thought and that for the sake of which the thought exists is the same; for you cannot find thought without something that is, as to which it is uttered. And there is not, and never shall be, anything besides what is, since fate has chained it so as to be whole and immovable. Wherefore all these things are but names which mortals have given, believing them to be true–coming into being and passing away, being and not being, change of place and alteration of bright colour.

Since, then, it has a furthest limit, it is complete on every side, like the mass of a rounded sphere, equally poised from the centre in every direction; for it cannot be greater or smaller in one place than in another. For there is no nothing that could keep it from reaching out equally, nor can aught that is be more here and less there than what is, since it is all inviolable. For the point from which it is equal in every direction tends equally to the limits.

We have in Parmenides' Poem a metaphysical exposition in which he argues for the eternal nature and necessity of *being* itself. Moreover, in saying, *"Nor is there nor will there be anything apart from being; for fate has linked it together, so that it is a whole and immovable."*, Parmenides argues that nothing exists apart from *being*, and that everything that *is* participates in the coherent consilient nature of unitary *being*.

Philosophical Arguments for the Existence of God

We can now review some of the main arguments for the existence of God, including a modified Argument from Truth. George Mavrodes provides a framework for categorization of various arguments (Mavrodes, 2005), stating that:

Most theistic arguments fall into one of two classes–the a priori or purely conceptual arguments, and the world-based arguments. The various versions of the ontological argument constitute the first class. . . .They have the advantage of concluding straightforwardly to the necessary existence of God, a feature which many take to be essential to the concept of a divine being.

In the other class belong the cosmological arguments, appealing to the general features of the world, and teleological [sic] arguments. . . .[A]nd there are some even more special arguments (perhaps versions of the teleological family)–arguments based on the demands of morality, the existence of beauty, the normativity of human rationality, religious experience, etc. . . .

There are other lines of argument which are not really intended to establish the truth of God's existence, but rather the rationality, the intellectual permissibility, etc., of theistic belief. Pascal's wager is an example, and the rather similar "will to believe" of William James [is another example].

With this framework in mind, we can begin to consider some of the main arguments.

In the Classical period, Plato set forth The Cosmological Argument in his *Timaeus* with its thesis of a demiurge and first cause. This argument was refined by Aristotle in his "Analytics, & Metaphysics"; after which it passed into Arab philosophy in the form of the Kalam argument by way of al-Kindi, Saadia al-Ghāzāli, Averroes, and then back to Western Europe via St. Bonaventure and

St. Thomas Aquinas (see below). A modern formulation of the argument is offered by William Lane Craig, (Craig 1979, 1980), in which he observes that the Kalam Cosmological Argument has two parts the first of which argues as follows:

1. Everything that begins to exist has a cause of its existence.
2. The universe began to exist.
3. Therefore, the universe has a cause of its existence.

The second part of the argument attempts to demonstrate that only a personal being could be the cause of a universe that has a beginning. From the perspective of modern cosmology, we would say a *spatio-temporal* beginning. In the event of an eternal multiverse that has neither beginning nor end, however, the cosmological argument fails because the premise of a beginning is false. This problem, and a resolution, will be discussed further when we review the various arguments of St. Thomas Aquinas below.

Augustine of Hippo [St. Augustine]

Concerning Augustine's argument for the existence of God in *De Libero Arbitrio*, John Mourant writes (Mourant 1971):

> In refutation of the skeptics, Augustine begins by establishing the possibility of knowledge, initiating a method that was to be followed more narrowly later by Descartes. Analyzing next the nature of knowledge, Augustine proceeds from a consideration of the knowledge attained by the senses and the function of such knowledge to a consideration of the function of the intellect and the kind of truths it attains. Describing the essential characteristics of both mathematics and moral truths he points out that they possess the characteristics of being eternal, immutable and necessary. From this the inference is then drawn that that which is eternal, immutable and necessary cannot be regarded either as being created by the mind of man or as existing innately within the mind of man. This entails that such truths are objective intelligible realities transcendent to, or "above" the human mind. As such they are dependent upon God and in some way may be identified with God Himself as eternal, necessary and immutable truth. In a word, the existence of such truths as immutable, necessary and eternal leads to the intuition that God exists as truth itself.

This is what has become known as *The Argument from Truth* about which Peter Kreeft and Ronald Tacelli were cited in the *Preface* as stating:

> And there is a good deal to be said for this. But that is just the problem. There is too much about the theory of knowledge that needs to be said before this could work as a persuasive demonstration.

There is perhaps a lack of total specificity in the argument concerning how "eternal, necessary, immutable" truth is "dependent upon God". We believe that if there is anything lacking in Augustine's Argument in this regard, it can be overcome by bringing the concepts of Consilient Truth and the epistemology of Dewey and Bentley into consideration. We will return to this after first examining what more Augustine and his successors had to say about the existence of God.

Mourant further observed that:

> ...it has been argued that there are elements of an Ontological Argument in the writings of Augustine.... A commentary on Psalm LXXII, 28, seems to suggest something like an ontological argument:

> Rational soul conceives God because it conceives that which is immutable and does not change. But both soul and body admit of change; thus that which is immutable is evidently superior to that which is not, and nothing is superior to the rational soul except God. If then it conceives something immutable, it is that without any doubt that it conceives.

> Parts of this argument seem close to Anselm without being explicitly the Anselmian argument.

If Augustine incompletely anticipates the Ontological Argument of Anselm, Boethius anticipates it quite closely.

Severinus Boethius [St. Boethius]

Boethius was born in 480, AD in Rome to a patrician family. Owing to his great learning, he was appointed to the service of Theodoric the Great. Theodoric the Great was king of the Germanic Ostrogoths (475–526), ruler of Italy (493–526), regent of the Visigoths (511–526), and was given the titles of Viceroy of the Eastern Roman Empire, Patrician, Consul of Rome, and Magister militum (master of the soldiers) by the Roman Emperor Zeno (Emperor of the Eastern Roman Empire). In effect, Theodoric was recognized legally and formally as the Western Roman Emperor and accorded all the formal titles by the Eastern Roman Emperor, even though he had achieved his status by means of barbaric conquest. Thus, the formalities of Rome continued on after 476 AD. The Ostrogoths eventually converted to Christianity (Gibbon 1776; Bury 1928; Moorhead 1992). Boethius fell under suspicion of treason and was executed by Theodoric in 524 AD. During his imprisonment, while awaiting his own execution and contemplating death, Boethius wrote one of the masterpieces of western philosophy, "The Consolation of Philosophy" (Boethius 1998). Here is part of Gibbon's (Ibid) entry on the last days of St. Boethius:

> While Boethius, oppressed with fetters, expected each moment in the tower of Pavia the sentence or the stroke of death, he composed The Consolation of Philosophy; a golden volume not unworthy of the leisure of Plato or Tully, but which claims incomparable merit from the barbarism of the times and the situation of the author. The celestial guide whom he had so long invoked at Rome and Athens now condescended to illumine his dungeon, to revive his courage and to pour into his wounds her salutary balm. ... From the earth Boethius ascended to Heaven in search of the SUPREME GOOD; explored the metaphysical labyrinth of chance and destiny; of prescience and free will; of time and eternity; and generously attempted to reconcile the perfect attributes of the Deity with the apparent disorders of his moral and physical government. ..... Suspense, the worst of evils, was at length determined by the ministers of death, who executed, and perhaps exceeded the inhumane mandate of Theodoric. A cord was fastened round the head of Boethius and forcibly tightened till his eyes almost started from their sockets; and some mercy may be discovered in the milder torture of beating him with clubs till he expired. But his genius survived to illumine a ray of knowledge over the darkest ages of the Latin world.

In Boethius' beautiful last work, *Philosophy* appears to him in a female persona and discourses with him regarding the necessity of a Supreme Good by virtue of this argument (Ibid. pp. 103–104):

> ...if a certain imperfection is visible in any class of things, it follows that there is also a degree of perfection in it. For if you do away with perfection it is impossible to imagine how that which is held to be imperfect could exist. The natural world did not take its origin from that which was impaired and incomplete, but issues from that which is unimpaired and perfect and then degenerates into this fallen and worn out condition.... since, there is a certain imperfect happiness in perishable good, there can be no doubt that a true and perfect happiness exists....

> As to where it can be found, you should think as follows... *Since nothing can be conceived better than God, everyone agrees that that which has no superior is good.* Reason shows that God is so good that we are convinced that His goodness is perfect. Otherwise.....There would have to be something else possessing perfect goodness over and above God, which would seem to be superior to Him and of greater antiquity. Therefore to avoid an unending argument, it must be admitted that the supreme God is to the highest degree filled with supreme and perfect goodness. [Emphasis added]

Philosophy further argues that God is the necessary source of perfect happiness and that God and perfect happiness are "one and the same thing", and that "God is to be found in goodness itself and nowhere else" (Ibid. p 108). This argument of Boethius was propounded in a different form later in the Middle Ages by Anselm of Canterbury, who argued for "that than which no *greater* can be thought" instead of "that which has no superior is good".

Anselm of Canterbury [St. Anselm]

The Ontological Argument has been most closely associated with St. Anselm of Canterbury (1033–1109 A.D.), canonized saint and scholastic medieval philosopher of the first rank St. Anselm, often called the Father of Scholasticism, was born in Aosta in the Kingdom of Burgundy. Today Aosta belongs to the Val d'Aosta region of Italy. St. Anselm later became Prior (1063) and then Abbot (1078), of the Monastery of Bec-Hellouin in Normandy, France. In 1093, he was consecrated Archbishop of Canterbury in England (Spade 2013). As an intellectual, he is known above all for three works (1) the "Monologion"; (2) the "Prosologion"; and (3) the "Cur Deus Homo". Known for his motto, *"fides quaerens intellectum"* ("Faith seeking understanding"), St. Anselm "was the outstanding Christian philosopher and theologian of the eleventh century [A.D.]." (Williams 2016).

It is interesting to observe how St. Anselm's ontology of God evolved in the course of his writings. To begin, Julius Weinberg paraphrases Anselm's argument in the Monologion as follows (Weinberg 1964, pp. 64–65):

> Whenever several things possess any attribute, whether in equal or unequal degree, they possess it by virtue of something that is the same in all of them. That by virtue of which several things possess the same attribute (in equal or unequal degree) must be that attribute in its highest degree and it must be that attribute existing through itself (and not through some other thing). Now since we see several degrees of goodness in things, it follows that there

must be something which is supremely good through itself. Again as we know that there are degrees of greatness, we can similarly infer that there is a maximum greatness, i.e., a being which is great through itself. Most important of all, we know that whatever exists exists by virtue of something or by virtue of nothing. But it is out of the question that something can exist by virtue of nothing; therefore whatever exists exists through (or by means of) something. Hence, either all things exist through many things which, in turn, exist through (or by means of) themselves, or there is exactly one thing existing through itself by means of which all other things exist. Now if there are several things each of which have the attribute of existing through itself, there must be something which IS this very attribute of existing through itself, for we cannot suppose that a plurality of self-existing things give to each other this attribute of self existence for this would mean that an independently existing thing derived its existence from another thing and so would be both independent and yet dependent. The only alternative remaining, then, is that there can be only one thing through which all other things exist. This Being is the maximum in the hierarchy of beings.

Williams (Ibid) cites Anselm's summary of his arguments in the "Stanford Encyclopedia of Philosophy" as follows:

Therefore, there is a certain nature or substance or essence who through himself is good and great and through himself is what he is; through whom exists whatever truly is good or great or anything at all; and who is the supreme good, the supreme great thing, the supreme being or subsistent, that is, supreme among all existing things.

This is reminiscent of Parmenides argument for the One Being that is the source of existence, and Boethius' argument for the Greatest Good as God.

Anselm was not satisfied with the arguments in the Monologion owing to their "complexity" and dependence on other assumptions. Weinberg (1964, p. 66) comments on this point citing Anselm's own words:

Anselm considered the arguments in the Monologion quite complicated,
    "The book was put together by connecting many arguments,"
    And so he asked himself whether he could discover a single argument (unum argumentum), which required no other proofs for its support, by which the existence of God can be demonstrated. And he reported that, after a long struggle to put the matter out of mind, the proof suddenly occurred to him.[1]

Weinberg (1964, p. 66) then summarizes Anselm's Ontological Argument of the "Proslogion":

Faith provides us with the conception of "that than which no greater can be conceived". Yet the Psalmist asserts, "The fool hath said in his heart there is no God" (Psalms 14:1). But the fool who hears of that than which no greater can be conceived understands what he hears. Now what we understand must at least exist in our understanding. But that than which nothing greater can be conceived cannot merely exist in the understanding, for if that than which no greater can be thought exists solely in the understanding it is not that than which no greater can be thought. For it is greater to exist in fact as well as in the mind than merely to exist in the mind; hence that than which no greater can be conceived exists both in the understanding and in reality. This Being exists so truly that it cannot be thought not to exist. Thus, if we can conceive of that than which no greater can be conceived, it must truly exist, for it is a contradiction to assert that that than which no greater can be conceived does not exist.

---

[1]Here, we have another interesting case of spontaneous insight in the mental efforts of a great medieval philosopher!

Kent summarizes Anselm's Ontological Argument succinctly (Kent W 1907):

Anselm's chief achievement in philosophy was the Ontological Argument for the existence of God put forth in his "Prosologium" [sic]. Starting from the notion that God is "that than which nothing greater can be thought", he argues that what exists in reality is greater than that which is only in the mind; wherefore, since "God is that than which nothing greater can be thought", [H]e [God] exists in reality. The validity of the argument was disputed at the outset by a monk named Gaunilo, who wrote a criticism on it, to which Anselm replied.

Williams (Ibid) provides more detail on Guanilo's criticism:

A monk named Gaunilo wrote a "Reply on Behalf of the Fool," contending that Anselm's argument gave the Psalmist's fool no good reason at all to believe that that than which a greater cannot be thought exists in reality. Gaunilo's most famous objection is an argument intended to be exactly parallel to Anselm's that generates an obviously absurd conclusion. Gaunilo proposes that instead of "that than which a greater cannot be thought" we consider "that island than which a greater cannot be thought." We understand what that expression means, so (following Anselm's reasoning) the greatest conceivable island exists in our understanding. But (again following Anselm's reasoning) that island must exist in reality as well; for if it did not, we could imagine a greater island–namely, one that existed in reality–and the greatest conceivable island would not be the greatest conceivable island after all. Surely, though, it is absurd to suppose that the greatest conceivable island actually exists in reality. Gaunilo concludes that Anselm's reasoning is fallacious.

Gaunilo's counterargument is so ingenious that it stands out as by far the most devastating criticism in his catalogue of Anselm's errors...

Correctly understood, Anselm says, the argument of the Proslogion can be summarized as follows:

1. That than which a greater cannot be thought can be thought.
2. If that than which a greater cannot be thought can be thought, it exists in reality. Therefore,
3. That than which a greater cannot be thought exists in reality.

Anselm defends .... by showing how we can form a conception of that than which a greater cannot be thought on the basis of our experience and understanding of those things than which a greater can be thought. For example,

it is clear to every reasonable mind that by raising our thoughts from lesser goods to greater goods, we are quite capable of forming an idea of that than which a greater cannot be thought on the basis of that than which a greater can be thought. Who, for example, is unable to think ... that if something that has a beginning and end is good, then something that has a beginning but never ceases to exist is much better? And that just as the latter is better than the former, so something that has neither beginning nor end is better still, even if it is always moving from the past through the present into the future? And that something that in no way needs or is compelled to change or move is far better even than that, whether any such thing exists in reality or not? Can such a thing not be thought? Can anything greater than this be thought? Or rather, is not this an example of forming an idea of that than which a greater cannot be thought on the basis of those things than which a greater can be thought? So there is in fact a way to form an idea of that than which a greater cannot be thought. (Anselm's Reply to Gaunilo 8)

Concerning the subsequent history of Anselm's Ontological argument, Kent (Ibid) states:

> ....this famous argument, which was lost and found again, [was] pulled to pieces and restored in the course of controversy.....it was revived in another form by Descartes [and later [revived in a modal form] by Leibniz]. After being assailed by Kant, it was defended by Hegel, for whom it had a peculiar fascination – he recurs to it in many parts of his writings. In one place he says that it is generally used by later philosophers, "yet always along with the other proofs, although it alone is the true one" (citation omitted)....if this proof were...an absurd fallacy, how could it appeal to such minds as those of [St.] Anselm, Descartes and Hegel? It may be well to add that the argument was not rejected by all the great Schoolmen. It was accepted by Alexander of Hales (citation omitted) and supported by [Duns] Scotus (citation omitted). In modern times it is accepted by Möhler, who quotes Hegel's defense with approval..."

The Five Arguments of Thomas Aquinas [St. Thomas]

Having covered the main ontological arguments, we can now turn our attention to those arguments that are best described as cosmological. The arguments of Thomas Aquinas fall into this category. According to Aquinas, the existence of God is "demonstrable and can be proven in five ways" ("*Quinque Viae.*"). The Five Arguments of Aquinas are (1) The Argument of the Unmoved Mover (2) The Argument of the First Cause (Efficient Causation) (3) The Argument of Contingency (4) The Argument from Degree and (5) The Teleological Argument. These are presented below in Thomas' own brief words (Aquinas 1947). Later we will consider the second and third arguments in particular:

> The existence of God can be proved in five ways. The first and more manifest way is the argument from motion. It is certain, and evident to our senses, that in the world some things are in motion. Now whatever is in motion is put in motion by another, for nothing can be in motion except it is in potentiality to that towards which it is in motion; whereas a thing moves inasmuch as it is in act. For motion is nothing else than the reduction of something from potentiality to actuality. But nothing can be reduced from potentiality to actuality, except by something in a state of actuality. Thus that which is actually hot, as fire, makes wood, which is potentially hot, to be actually hot, and thereby moves and changes it. Now it is not possible that the same thing should be at once in actuality and potentiality in the same respect, but only in different respects. For what is actually hot cannot simultaneously be potentially hot; but it is simultaneously potentially cold. It is therefore impossible that in the same respect and in the same way a thing should be both mover and moved, i.e. that it should move itself. Therefore, whatever is in motion must be put in motion by another. If that by which it is put in motion be itself put in motion, then this also must needs be put in motion by another, and that by another again. But this cannot go on to infinity, because then there would be no first mover, and, consequently, no other mover; seeing that subsequent movers move only inasmuch as they are put in motion by the first mover; as the staff moves only because it is put in motion by the hand. Therefore it is necessary to arrive at a first mover, put in motion by no other; and this everyone understands to be God.

> The second way is from the nature of the efficient cause. In the world of sense we find there is an order of efficient causes. There is no case known (neither is it, indeed, possible) in which a thing is found to be the efficient cause of itself; for so it would be prior to itself,

which is impossible. Now in efficient causes it is not possible to go on to infinity, because in all efficient causes following in order, the first is the cause of the intermediate cause, and the intermediate is the cause of the ultimate cause, whether the intermediate cause be several, or only one. Now to take away the cause is to take away the effect. Therefore, if there be no first cause among efficient causes, there will be no ultimate, nor any intermediate cause. But if in efficient causes it is possible to go on to infinity, there will be no first efficient cause, neither will there be an ultimate effect, nor any intermediate efficient causes; all of which is plainly false. Therefore it is necessary to admit a first efficient cause, to which everyone gives the name of God.

The third way is taken from possibility and necessity, and runs thus. We find in nature things that are possible to be and not to be, since they are found to be generated, and to corrupt, and consequently, they are possible to be and not to be. But it is impossible for these always to exist, for that which is possible not to be at some time is not. Therefore, if everything is possible not to be, then at one time there could have been nothing in existence. Now if this were true, even now there would be nothing in existence, because that which does not exist only begins to exist by something already existing. Therefore, if at one time nothing was in existence, it would have been impossible for anything to have begun to exist; and thus even now nothing would be in existence---which is absurd. Therefore, not all beings are merely possible, but there must exist something the existence of which is necessary. But every necessary thing either has its necessity caused by another, or not. Now it is impossible to go on to infinity in necessary things which have their necessity caused by another, as has been already proved in regard to efficient causes. Therefore we cannot but postulate the existence of some being having of itself its own necessity, and not receiving it from another, but rather causing in others their necessity. This all men speak of as God.

The fourth way is taken from the gradation to be found in things. Among beings there are some more and some less good, true, noble and the like. But "more" and "less" are predicated of different things, according as they resemble in their different ways something which is the maximum, as a thing is said to be hotter according as it more nearly resembles that which is hottest; so that there is something which is truest, something best, something noblest and, consequently, something which is uttermost being; for those things that are greatest in truth are greatest in being, as it is written in Metaph. ii. Now the maximum in any genus is the cause of all in that genus; as fire, which is the maximum heat, is the cause of all hot things. Therefore there must also be something which is to all beings the cause of their being, goodness, and every other perfection; and this we call God.

The fifth way is taken from the governance of the world. We see that things which lack intelligence, such as natural bodies, act for an end, and this is evident from their acting always, or nearly always, in the same way, so as to obtain the best result. Hence it is plain that not fortuitously, but designedly, do they achieve their end. Now whatever lacks intelligence cannot move towards an end, unless it be directed by some being endowed with knowledge and intelligence; as the arrow is shot to its mark by the archer. Therefore some intelligent being exists by whom all natural things are directed to their end; and this being we call God.

The Argument from *Efficient Causation*, or "The Argument of the First Cause", is the best known. The stage for Aquinas to make this argument was set by the aforementioned Cosmological Argument of Plato: *everything that begins to exist has a cause*. Taking this lead, Aquinas argues that all things observed can be seen to have a cause, which is by definition antecedent to its effect. This leads immediately to the concept of the cause of the cause of the effect, which in turn has its own cause.

A regressive infinity of causes would be required if there were not a first cause that is the sufficient cause of itself. This First Cause is God.[2]

The Third of Aquinas' Arguments will also be discussed briefly here. The *Argument from Possibility* (Argument from Contingency, "*ex Contingentia*") is summarized by Weinberg (1964, p. 66) as follows:

> We observe that some things in nature come into existence and pass away after a time. From this we can immediately conclude that such things are merely possible beings in the sense that they can either be or not be. And that which is possible must at some time come to an end, assuming that it exists. If this had happened to all such possible beings a moment would ultimately arrive at which nothing of the sort exists. But then, since all these beings were merely possible, nothing could come into existence. Now if all beings were merely possible nothing would exist, and so we must conclude that not all beings are [merely] possible, i.e. that there is a necessary being. There cannot be a sequence of necessary beings which owe their existence to other necessary beings ad infinitum; hence, there must be a necessary being which is unqualifiedly necessary, i.e., a necessary being that is the cause of all being.

Together, the arguments from Efficient Causation and Possibility/Contingency lead to a Necessary Being that is the self-sufficient First Cause of all that exists. These arguments, however, require adaptation to a multiverse cosmology in the event that the multiverse is proven to be real. In the context of a multiverse cosmology, we need to distinguish between two possible cases. The first is an *infinite* multiverse, ie. a multiverse that has a beginning but does not end; and the second is the case of an *eternal* multiverse that has neither beginning nor end. In the case of an infinite multiverse, it is easier to admit the validity of the Argument from Efficient Causation, because causation is consistent with having a beginning. The Argument from Possibility may also be valid in this case, because anything that has a beginning cannot lay claim to necessity of being. In this case, the infinite multiverse would consist of a sequence of possible realities that according to the Argument from Possibility requires a Necessary Being as its cause.

The case of an eternal multiverse, that has neither beginning nor end, is more problematic for Aquinas' arguments, however. One could argue in the case of an eternal multiverse that because it has no beginning the Argument from Efficient Causation does not apply, or that it is the wrong causal argument because it views causation from a time-bound perspective when it argues for a *first* cause rather than a transcendental cause that affects all spatio-temporal domains simultaneously. We might then argue that the correct causal argument is one in which the Necessary Being exists in a higher dimensional or transcendental Archimedean domain from where He causes all else to come into being in lower dimensional space-time. In the eternal multiverse scenario, the necessary higher dimensional Being and the

---

[2]The Argument from Efficient Causation is closely related to the Principle of Sufficient Reason ("PSR"), which articulates (1) For every entity X, if X exists, then there is a sufficient explanation for why X exists; (2) For every event E, if E occurs, then there is a sufficient explanation for why E occurs; (3) For every proposition P, if P is true, then there is a sufficient explanation for why P is true. The PSR was advanced by the ancient Greeks including the pre-Socratics, but found its most formal expression in the works of Spinoza, Leibniz and especially Arthur Schopenhauer, who wrote a famous doctoral thesis in 1813 "On the Fourfold Root of the Principle of Sufficient Reason".

multiverse *In Toto* must be co-eternal, while each of the multiverse's component universes has a definitive beginning and a potential ending in space-time.

On the other hand, one could also argue that an *eternal* multiverse, as defined, is necessary in its own right, because *In Toto* it has neither beginning nor end. In this case, the possibility that it is the sufficient cause of itself must be considered. Could we conclude that an eternal multiverse can be the *necessary being which is unqualifiedly necessary, i.e., a necessary being that is the cause of all being, which is to say the cause of itself?* Can this argument be sustained? The eternal multiverse envisioned in modern cosmology is postulated to consist of an infinite number of pocket universes (or finite domains) each of which has its cause and beginning from an antecedent one. This sequence of pocket universes extends in both directions of space-time and has neither beginning nor end, so collectively an eternal multiverse may be characterized as having necessity of being, but none of its constituent parts has necessity of being. In this argument, the *necessary* eternal multiverse consists exclusively of an eternal sequence of merely *possible or contingent* pocket universes. Is it reasonable to conclude that a multiverse that embodies such a contradiction is the sufficient cause of itself? We must answer *no* because as quoted above according to the Argument from Contingency:

> Now if all beings were merely possible nothing would exist, and so we must conclude that not all beings are [merely] possible, i.e. that there is a necessary being.

It is more reasonable to conclude that the Unqualified Necessary Being does not exist in space-time at all, but rather in a higher dimensional transcendental realm that is the "point of origin"[3] and causation for the eternal sequence of pocket universes that comprises the eternal multiverse. That is, the eternal multiverse is co-eternal with the Unqualified Necessary Being. So once again, in the case of an eternal multiverse we see that the argument from Efficient Causation and the Argument from Contingency lead to a Necessary Being that is the self-sufficient Cause of all that exists. We are warranted, on the basis of this conclusion, to observe that the *Necessary Being that is the self-sufficient Cause of all that exists* is the Supreme Being Who St. Anselm defined as, "that than which no greater can be thought"!

We have seen in the preceding sections how St. Anselm's Ontological Argument, and St. Thomas's Arguments from Efficient Causation and from Contingency create a strong case in support of the existence of God, even when adapted to the case of an eternal multiverse. For either there is a necessary cause of the multiverse that exists apart from or outside of it, or the multiverse is the sufficient cause of itself. The Argument from Contingency demonstrates why the universe, or even an eternal multiverse, cannot be the sufficient cause of itself. Nevertheless, Physicist, Lawrence Krauss argues a scientific hypothesis for a universe from nothing in his book of the same title (Krauss 2012). Mathematician, Amir Aczel, provides a cogent rebuttal in which he argues for the necessity of a pre-existent quantum foam, or reality, from

---

[3]The Archimedean Point of Huw Price alludes to the perspective of God in viewing the universe. John Bell also invoked this point of view when he spoke of "something coming from outside the universe" to explain the instantaneous communication between entangled quantum particles.

which a singularity gave birth to our universe (Aczel 2014). This is consistent with the theory of eternal inflation that gives rise to a self-replicating fractal multiverse as discussed in Chap. 4. Moreover, Krauss fails to explain how the laws of physics came into being.

Despite the scientific and philosophical arguments that the multiverse cannot be the sufficient cause of itself, atheists counter with the same question in regard to a Supreme Intelligence or Being. Essentially this question is heard in the form of, "if God created the Universe then who created God". Certainly, an answer may be found in Anselm's reasoning that a created god is not "that than which no greater can be thought" in conjunction with the Aquinan arguments for the necessity of a self-sufficient being that is the First Cause of all that exists. Another, and perhaps more cogent, response approaches the objection on the basis of St. Augustine's Argument from Truth, which is an epistemological argument. The modification of St. Augustine's Argument from Truth that is presented below argues for the existence of a Supreme Intelligent Being Who is the necessary *Knower* of Eternal Consilient Truth.

Recollection and Synthesis

Much ground has been covered, and we have hit upon some vital clues in earlier chapters. To proceed further it will be helpful to recall some of those key observations. It is also worth reminding ourselves that, while many diverse fields of inquiry have been touched upon, the valid findings of each are a part of the unitary and coherent entity that E.O. Wilson described in his book, "Consilience: The Unity of Knowledge". We have proceeded, therefore, with the strong conviction that regardless of how unrelated the statements that arise from various methods of inquiry may appear, all truths apprehended are part of a grand coherence by virtue of consilience. This connectedness applies not only to all subsidiary truths, but also to the ultimate truth to which they are subordinate or from which they arise. For this reason, we have consistently argued for a synthesis of information derived from theological, philosophical and scientific methods of inquiry. A corollary to the notion that true statements discovered by these different disciplines must not be contradictory is the concept that much of the debate between proponents of science and religion is based upon false arguments that are irrelevant to the essential ontological question relating to God's existence.

About the Universe-Multiverse

1. The universe that constitutes the reality in which we dwell had a definitive beginning about 13.5 billion years ago.
2. It consists of matter and energy that is manifested in the elementary particles and their associated fields at the quantum level.

3. The universe is characterized by (a) precisely defined physical constants that appear to have values that are consistent with the emergence of life and mind from inanimate precursors, and (b) physical laws, expressed in the universal language of mathematics, that govern the interactions of matter and energy over both atomic and cosmic scales.

4. Quantum reality is fraught with paradox, which implies that there is a deeper level of reality that we do not yet understand.

5. The universe may be only one component of what has been called the multiverse, which may in turn be an infinite or eternal system of pocket universes that exist sequentially in space-time, with perhaps more than one existing in parallel "at the same time". This multiverse is infinite at a minimum and perhaps even eternal. In any case, it is vast and the truth that can be known about it exceeds the span of apprehension of even a very large number of finite minds. This can be said about our own pocket universe as well.

About Logic

1. Truth is *that which can be proved and known* without contradiction.

2. The *apprehension* of Truth is imperfect, however, because logic depends on the truth of assumptions or predicate statements, and because experiment is subject to numerous sources of error and is limited by the scope of hypotheses tested. "Truth" thus apprehended may subsequently be shown to be false by superior hypotheses, methods and experiments.

3. It follows that mutable, finite mind, can discover or know certain and/or absolute truth about the objects of mathematics and logic only insofar as the predicates or axioms of proof are certain and the methods and the laws of logic are applied without error.

4. Likewise, truth about the objects of reality cannot be known by humanity with certainty owing to the limitations of empirical methods and the defined scope of inquiry. Nevertheless, what can be known about the quantum objects of reality is assumed to be defined by the quantum mechanical wave function that describes those objects.

5. A set is a collection of tangible or intangible objects that is defined by some criterion of inclusion.

6. The set of all true statements about mathematics and the universe or multiverse is a vast, and possibly infinite, set. Each of the vast, and potentially infinite, number of quantum particles has a quantum wave function that describes the probabilities of its location throughout all of space-time. Its wave function is what can be asserted truly or known about a quantum particle.

7. This potentially infinite set of true mathematical and logical statements, together with the true statements about the multiverse, is referred to as Consilient Truth.

8. Consilient truth is therefore what can be discovered or *known* about logic, mathematics and the multiverse collectively. It is not the objects of logic, mathematics or the multiverse, *per se*.
9. Consilience is the term that refers to the internal coherence and unitary nature of the elements of Consilient Truth.

About the Ontological Question

1. The cause of the multiverse is unknown, but religion postulates that it is the Supreme Intelligence called God.
2. A definitive proof for, or against, the existence of God is disputed.

The itemization set forth above is not presented as axiomatic or in a dogmatic fashion, because what is known about mathematics, the universe-multiverse and logic is subject to further inquiry and modification. It seems reasonable, however, to take note of these items as reflecting a reasonable approximation to the current consensus view among scientists, mathematicians, philosophers and theologians.

With this background in mind, we can attempt a modified argument from truth for the existence of God, as follows.

The Modified Argument from Truth

1. *Truth is what can be proved and therefore known without contradiction* about the objects and operations of logic, mathematics and the multiverse. Collectively, the true statements about the infinite number of objects of the multiverse, logic and mathematics are defined as *Consilient Truth. It is not the objects or operations themselves, nor does it have its existence in the objects or operations.*
2. Every element of Consilient Truth can be deduced in principle by some sequence of deductive logical steps from any other element, because true statements are never contradictory. The true statements which comprise Consilient Truth are, therefore, necessarily true, and *they exist necessarily.*
3. The true statements of necessary Consilient Truth do not have their existence in the objects to which they refer (see 1). Therefore, they must exist transcendentally apart from the space and time of the multiverse.
4. Necessary Consilient Truth is therefore eternal.
5. Knowledge, *per se*, has no existence apart from a knowing mind. There is no transcendental repository for knowledge outside of mind, such as that postulated by Plato in his concept of a Realm of Ideas, Ideals, or Forms.
6. Only the *act of knowing* and its object, that which is *known,* have epistemological and ontological validity.

7. The necessity of Consilient Truth's existence may be demonstrated deductively as above, but the actuality and the incidentals of its existence derive from being known by mind. That is, Consilient Truth subsists in mind and only in mind.
8. Necessary, transcendental, eternal Consilient Truth cannot be known eternally by mutable finite mind. It must therefore be known by, and have its eternal existence only in the active eternal knowing by, a necessary, eternal and transcendental mind.
9. That mind is the mind of God. Therefore, God exists necessarily, transcendentally, and eternally which was to be proved.

This proof hinges upon a few basic and simple statements. These are: (1) Consilient Truth is necessary, transcendental and therefore eternal; (2) Consilient Truth has its only existence in knowing mind; (3) That mind must, therefore, be necessary, transcendental, and eternal; and (4) such a mind must be possessed by a Being commonly referred to as God, Who exists necessarily, transcendentally and eternally.

Ask yourself if you believe that necessary truth can exist apart from knowing mind. If you answer, yes, then ask yourself in what it exists. In what does it have its being, or in what does it subsist? Earlier, we referred to Consilient Truth as the *set of true statements*. We emphasize here, that *set of true statements* is a cognitive construct, that is used to define Consilient Truth and has its existence necessarily in mind because it is an idea. In contrast, even though Plato used the language of cognition to describe the *Realm of Ideas* (Ideals or Forms), he was referring to an actual reality that contains an infinite transcendental compendium of truth that exists outside of mind. Consider for a moment the irony of *believing* that such an infinite *Realm of Ideas* can have its own existence apart from a knowing mind. If you still believe this is true, then you are effectively rejecting Dewey's and Bentley's argument against the existence of knowledge *per se*, and you are embracing the idea of an ethereal transcendental, infinite and eternal repository of information. We would submit to you that Plato's *Realm of Ideas*, as a repository of Consilient Truth, is even more fantastic than the idea of an Intelligent Supreme Being that *knows* Consilient Truth in its totality. Moreover, a *mindless* repository of all truth presents several serious difficulties. Who or what created this repository, and how does it have its being? Further, an infinite repository of Consilient Truth would be without meaning or effect if it is not understood by mind! It is rather like a book that is never read. Truth without meaning or effect is paradoxical. It is a logical contradiction.[4] It is more parsimonious to acknowledge the existence of an eternal transcendent mind that does the knowing, and that Consilient Truth has its only existence in such a mind. If you do accept the idea that Consilient Truth is eternal and beyond the knowing of which all the sapient minds in the universe (multiverse) are capable, and you accept the Dewey-Bentley philosophical argument that knowledge, per se, has

---

[4]Truth without meaning is an absurdity!

no existence apart from mind, then you must accept the necessity of God's existence as the Being that apprehends Consilient Truth.

We may conclude, therefore, that Consilient Truth exists and has its being in the mind of God. This would include all of physical law and the physical constants, together with mathematics as the universal language in which God knows and understands the foundations of all natural law and reality. What is the effect of this Divine knowing? We propose that the effect of the active Divine knowing of Consilient Truth is manifested in the continuous and eternal emanation of the Multiverse from the Mind of God into existence in space-time. The Multiverse must therefore be co-eternal with God.

Along these lines, we recall physicist Bernard Haisch's (Ibid) suggestion that God embodies *creative potential* that is actualized or manifested as the dynamic reality we observe in this universe. David Birnbaum also proposes the idea that God is the source of existence that is eternally actualized in every aspect of the dynamic universe (Birnbaum 2005). Writing from a Jewish philosophical and theological, yet cosmological, point of view Birnbaum cites from Encyclopaedia Judaica, in regard to the founder of Hasidic Judaism, Rabbi Yisroel ben Eliezer, who was also known as "The Baal Shem Tov" and "Besht", as follows:

> The foundation stone of Hasidism as laid by Besht is a strongly marked pantheistic conception of God. He declared the whole universe, mind and matter, to be a manifestation of the Divine being; that this manifestation is not an emanation from God, as is the conception of Kabala, for nothing can be separated from God: *all things are rather forms in which He reveals Himself.* [Emphasis Added]

Later, Birnbaum (Ibid) writes concerning God's action as emanating from the cosmic womb of potential as an *infinitely expanding* Being striving toward His own maximum expression:

> The Cosmic Womb of Potential nourishes, enhances, and sustains the potentializing of all aspects of the Cosmic order – including itself. . .
>
> The infinitely expanding and traversing aspect stretches forward through time. . . .from the very origin of origins to the forward reaches of time. This infinitely expanding dynamic/ entity strives after its own maximal potential, relentlessly advancing and expanding, again, recursively, a womb woven within a womb within a womb.

Spinoza's concept of God, which Einstein found to be of such interest, is also important in this context (Spinoza 1677. p. 122). Spinoza states that God is:

> Being absolutely infinite, that is to say, substance consisting of infinite attributes each one of which expresses eternal and infinite essence. . .

Concerning substance Spinoza (1677. p. 124) writes:

> By substance, I understand that which is in itself and is conceived through itself; in other words, that, the conception of which does not need the conception of another thing from which it must be formed."

Finally, in regard to the nature of God Spinoza (Spinoza 1677. p. 124) states:

> From the necessity of the divine nature infinite numbers of things in infinite ways (that is to say, all things which can be conceived by the infinite intellect) must follow.

Spinoza clearly speaks of God as a Being who is the source of all being; a Being who possesses a *mind* of infinite capability for ideation, and creation that necessarily follows from it. From this perspective, an eternally unfolding multiverse in space-time may be viewed, as the eternal expression of God's potential, thought or *idea*. That the laws of physics lead to the emergence of intelligent mind in the universe offers a beautiful symmetry to contemplate. With Mind God created the universe[5]; and mind arises in that universe, to search for the ultimate meaning and cause of existence.

As explained in Chap. 4, the eternal, self-similar, self-replicating multiverse has been described as a fractal geometric structure. This idea is captured implicitly in the quotes above from Birnbaum and Spinoza. It is perhaps more than a passing curiosity that the form in which God reveals Himself to Moses in Exodus is a burning bush that is not consumed.[6] Studies of the dynamic shape of a flame front show that it too is a fractal geometric structure (Troiani and Marrocco 2011). That the bush was not consumed by the fire conveys the idea that the source of God's creative action in the multiverse is inexhaustible and unending. The creative activity manifests as the eternal emanation of the multiverse in space-time. God's revelation of His name to Moses, *I Will Be Who I Will Be*, conveys the same meaning as the fractal geometry of the flame in which He appears: God's creation is an eternal emanation of infinite unending variety. In the eternal creative relationship between a transcendental God and the co-eternal Multiverse of space-time, we glimpse what St. Bernard (Bernard of Clairvaux, 2016) meant when he was inspired to write:

> God has spoken one time; certainly only one time, because always. For His speech is one and constant, continuous and perpetual.[7]

References

Aczel AD (2014) Why science does not disprove god. Harper Collins, New York

Aquinas T St (1947) Summa theologica. Benziger Bros. edition Translated by Fathers of the English Dominican Province. First Part: Question 2. URL: http://dhspriory.org/thomas/summa/FP/FP002.html#FPQ2OUTP1

Bernard of Clairvaux (2016) Monastic sermons. Cistercian Publications, Athens, p 35

Birnbaum D (2005) Summa metaphysica II: god and good. Harvard New Paradigm, Boston/New York, p 51

Boethius (1998) The consolation of philosophy. VE Watts (Ed.,transl). The Folio Society, London

Burnet J (1920) Parmenides on nature. Early greek philosophy, 3rd edn. A & C Black, London, pp 117–121

---

[5]Conveying an equivalent meaning, Gerald Schroeder wrote, "Wisdom or mind is the substrate of all creation". (Personal Communication)

[6]"And the Lord's messenger appeared to him in a flame of fire from the midst of the bush, and he saw, and look, the bush was burning with fire and the bush was not consumed". Exodus 3:2

[7]This quote is taken from St. Bernard commenting on Psalm 61:12 – "God has spoken once". An alternative rendering from the "Liturgy of the Hours" is: "God has spoken once. Once indeed, because forever. His is a single, uninterrupted utterance, because it is continuous and unending".

Bury JB (1928) The invasion of Europe by the barbarians. Macmillan, London

Craig WL (1979) The Kalām cosmological argument. Macmillan, London

Craig WL (1980) The cosmological argument from Plato to Leibniz. Macmillan, London

Gibbon E (1776–1789) The history of the decline and fall of the Roman empire. 5 pp 29–54. The Folio Society, London. 1987

Kent W (1907) "St. Anselm." The catholic encyclopedia. Vol. 1. Robert Appleton Company, New York. 9 Sept 2016. URL: http://www.newadvent.org/cathen/01546a.htm

Krauss LA (2012) Universe from nothing: why there is something rather than nothing. New York Free Press, New York

Mavrodes G. (2005) God, arguments for the existence of. The Oxford guide to philosophy, 2d edn (Honderich T ed). Oxford University Press, Oxford, p 343 & *passim*

Moorhead J (1992) Theodoric in Italy. Oxford University Press, Oxford

Mourant JA (1971) The augustinian argument for the existence of god. In: Ross JF (ed) Inquiries into Medieval Philosophy. Greenwood Publishing Co, Westport

Palmer J (2016) Parmenides, The Stanford Encyclopedia of Philosophy (Winer Edition), Zalta EN (ed) URL: http://plato.stanford.edu/archives/fall2016/entries/parmenides/

Spade PV (2013) Medieval philosophy. The Stanford encyclopedia of philosophy, Spring 2013 edn, Zalta EN (ed), URL: http://plato.stanford.edu/archives/spr2013/entries/medieval-philosophy/

Spinoza B (1677) Quoted in, The philosophy of Spinoza. The Modern Library, New York

The Stanford Encyclopedia of Philosophy. 2012 entry on parmenides

Troiani G, Marrocco M (2011) Imaging of flame fronts by fluorescence of hydroxyl radicals: a fractal approach for the determination of front position. XXXIV meeting of the Italian section of the combustion institute pp 1–8

Weinberg JR (1964) A short history of medieval philosophy. Princeton University Press, Princeton

Williams T (2016) Saint Anselm. In: Zalta EN (ed), The Stanford encyclopedia of philosophy, Spring 2016 edn. URL: http://plato.stanford.edu/archives/spr2016/entries/anselm/

Index

© Springer Nature Switzerland AG 2018
R. J. Di Rocco, *Consilience, Truth and the Mind of God*,
https://doi.org/10.1007/978-3-030-01869-6

Printed by Printforce, the Netherlands